**Books are to be returned on or before
the last date below.**

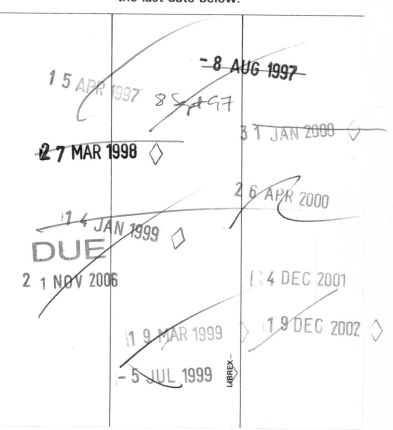

Electronics, Noise and Signal Recovery

MICROELECTRONICS AND SIGNAL PROCESSING

Series editor: **P.G. Farrell**, University of Manchester, U.K.

About this series:
The topic of microelectronics, the fastest growing subject in the engineering field, can no longer be treated in isolation from its prime application in the processing of all information-bearing signals. The relative importance of various processing functions will determine the future course of developments in microelectronics. Many signal processing concepts, from data manipulation to mathematical operations such as correlation, convolution and Fourier transformation, are now readily realizable in microelectronic form. This series aims to satisfy a demand for comprehensive and immediately useful volumes linking the microelectronic technology and its applications.

Key features of the series are:
● Coverage ranging from the basic semiconductor processing of microelectronic circuits to developments in microprocessor systems or VLSI architecture and the newest techniques in image, voice and optical signal processing.
● Emphasis on technology, with a blend of theory and practice intended for a wide readership.
● Exposition of the fundamental theme of signal processing; namely, any aspect of what happens to an electronic, acoustic or optical signal between the basic sensor which gathers it and the final output interface to the user.

Electronics, Noise and Signal Recovery

E.R. DAVIES

Department of Physics
Royal Holloway and Bedford New College (University of London)
Egham, Surrey, UK

ACADEMIC PRESS
Harcourt Brace & Company, Publishers
London   San Diego   New York
Boston   Sydney   Tokyo   Toronto

ACADEMIC PRESS LIMITED
24/28 Oval Road,
London NW1 7DX

United States Edition published by
ACADEMIC PRESS INC.
San Diego, CA 92101

ISBN 0-12-206130-6
0-12-206131-4 (pbk)

Typeset by P&R Typesetters Ltd, Salisbury, UK
Printed in Great Britain by TJ Press Ltd, Padstow, Cornwall

Preface

Electronics is the life-blood of modern society, having made its way not only into such common products as radios, light dimmers, microwave ovens, electronic ignition systems, and computers, but also into industrial control systems and earth-orbiting satellites – to name but a few applications. In addition, it is of vital importance in the laboratory in subjects as diverse as astronomy, physics, engineering, chemistry, biology, psychology, and many branches of medicine such as cardiology, radiology and tomography. In many of these cases, electronics is needed to extract tiny signals (often originating from small numbers of atoms) from a background of thermal fluctuations and other noise.

This book evolved over a period of some 20 years, from material I have given on undergraduate and postgraduate courses at London University. In addition, it has drawn on the practical experience I gained in my early investigations in solid state physics on magnetic resonance phenomena, with a further quite large impetus from my more recent work on computer vision. In both of these cases, the fight to extract small signals from noise* – a topic known as signal recovery – is intensive, and this book has aimed to include my experience in these and related areas within a coherent body of knowledge.

In signal recovery it is insufficient simply to amplify the small incoming signals to the required output level, since to do so would yield an output that is swamped by noise. Thus noise removal must be regarded as a central part of the process of signal recovery. To cope properly with this subject, some preparation on the topic of noise is clearly required, together with a reasonably thorough coverage of basic electronics. It therefore seemed natural to divide the book into three sections – Part 1 being devoted to *Electronics*, Part 2 to *Noise* and Part 3 to *Signal Recovery*, though some care has been taken to provide strong links and cross-references between the material in the various chapters.

---

*In vision, as in radar, noise arises not only as thermal fluctuations and interference but also in the form of currently irrelevant signals or "clutter".

As always, there is the problem of what can be included within reasonable space. The difficult decision was made to omit digital electronics, on the basis that the concepts of signal recovery turn out to be mainly analogue in nature (for a fuller justification of this see Chapter 21). Thus Part 1 is involved exclusively with analogue electronics. Similarly, computer algorithms have on the whole been omitted for much the same reason, though various chapters in Part 3 make it clear at exactly what stage this becomes a limitation. Specifically, a book on signal *processing* would not be able to exclude computer algorithms, whereas one on signal *recovery* can sensibly talk about various aspects of signal averaging and noise rejection, dealing with basic principles and basic electronic means of carrying out these processes, while demonstrating clearly the limitations which will make it vital to use computer methods. Nowhere is this clearer than in the field of computer vision, a topic which is introduced and covered carefully at an elementary level in Chapter 20.

But why restrict the discussion to signal recovery? The reason is the paucity of well-written reference material in this area. Previous monographs have tended to cover very restricted aspects of signal recovery, e.g. phase-sensitive detectors, or matched filters for radar. This book aims to bring together under a single heading the common aspects of signal recovery – frequency-domain filters, phase-sensitive detectors, boxcar detectors, signal averagers, matched filters, non-linear filters, and so on – and to treat them coherently. At the same time, it is insufficient in such a complex subject to deal with these topics in theory alone. Therefore, case studies have been introduced to illustrate the principles in fair depth, showing in particular how systems can be set up for optimum signal-to-noise ratio (see especially Chapters 13, 15, 17 and 20). The case studies should help the student or the practising experimental scientist or engineer to set up experiments or build equipment that is suitable for his application. In fact it is intended that the book will be useful not only for undergraduates in various scientific and engineering disciplines, but also for postgraduates and other researchers who have to build real equipment and who need solid help in how to design optimal systems. Finally, so that the book did not become too involved with the technical detail of specific applications, a bibliography section has been included at the end of each chapter, together with a considerable number of references.

Thus Part 1 should be suitable for science students in the second or third years, Part 2 should provide useful background on noise sources at third year level, while Part 3 should be relevant both for second and third year undergraduates and for postgraduates and practitioners in the field. Since the more advanced topics of Part 3 were the ultimate target of the book, it did not seem appropriate to extend the length further and include topics such as basic electricity, a.c. theory, electromagnetism, or semiconductor physics – it is assumed that these topics have been covered in a preliminary course from which this book can take over at whatever juncture seems appropriate. (However, an appendix summarizing relevant properties of semiconductors *is* included.)

As mentioned above, the book leans heavily on experience I have gained both in experimental solid state physics and in computer vision. Of particular note is the help I gained as a first year postgraduate from John Hurrell, who always seemed able to squeeze that extra decibel of signal-to-noise ratio out of a microwave bridge! Special thanks are due to my other magnetic resonance colleagues Michael Baker and John Davies, who also helped in various ways to shape my view of the subject. In addition, it is a pleasure to record many useful discussions with Brian Cowan and Stuart Flockton on the respective topics of noise and signal processing. Next, I am indebted to Andrew Carrick, of Academic Press, without whose help and patience this book might never have been published. Finally, it would be a gross oversight if I did not acknowledge the immense amount of support and encouragement I have received from my wife and children over rather many hours of writing!

Royal Holloway and Bedford New College,   E.R. DAVIES
University of London

Acknowledgements

I am indebted to the following who have offered a variety of constructive comments on portions of the manuscript:

J. Butt, B.P. Cowan, J.J. Davies, S.J. Flockton, G.A. Gledhill, A.I.C. Johnstone, M.J. Lea, C. Nicholls, M. van Daalen.

I am pleased to thank Dr J.M. Baker, Dr J.P. Hurrell and The Royal Society for permission to reproduce Figure 13.18(a) from the paper by Baker, Davies and Hurrell (1968) *Proc. Roy. Soc.*, **A308**, pp. 403–431.

Finally, I am pleased to thank the University of London for permission to reproduce parts of the following BSc examination questions as problems 11.1, 15.2, and 19.1 respectively:

Qu. 3 in EDP *Data Transmission and Processing*, RHC (1978)
Qu. 4 in PH281 *Analogue Electronics*, RHBNC (1992)
Qu. 3 in P310 *Data Transmission and Processing*, RHC (1976).

Contents

Part 2   Noise

Part 3   Signal Recovery

Appendices

About the Author

Dr E.R. Davies graduated from Oxford University in 1963 with a First Class Honours degree in Physics. During his 12 years' research in solid state physics investigating magnetic resonance phenomena and associated signal recovery methods, he developed the spectroscopic technique now widely known as *Davies-ENDOR*. Later, he became interested in vision and is now Reader in Machine Vision at Royal Holloway and Bedford New College, University of London. His current research is on noise suppression and robust pattern matching in digital images; he also specializes in algorithm design for automated visual inspection. Dr Davies is a Fellow of the Institute of Physics and of the Institution of Electrical Engineers, and is a member of the British Machine Vision Association and of the IEEE and IEEE Computer Society.

Dedication

This book is dedicated to those who have taught me, those who have inspired me, those who have helped me to learn for myself, and not least

- my teachers at School and College
- my colleagues at the Clarendon Laboratory and at London University
- my parents who started me off and encouraged me to be both scientist and engineer.

Glossary of Acronyms and Abbreviations

a.c.	alternating current
ADC	analogue-to-digital converter
a.f.	audio frequency
a.m.	amplitude modulation
BJT	bipolar junction transistor
CAT	computer of average transients
CB	common base
CC	common collector
CCD	charge-coupled device
CE	common emitter
CMRR	common mode rejection ratio
c.w.	continuous wave
DAC	digital-to-analogue converter
d.c.	direct current
ENDOR	electron nuclear double resonance
EPI	echo-planar imaging
EPR	electron paramagnetic resonance
FET	field-effect transistor
FID	free induction decay
f.m.	frequency modulation
GRE	gradient-recalled echo
HEMT	high electron mobility transistor
h.f.	high frequency
i.c.	integrated circuit
i.f.	intermediate frequency
IGFET	insulated gate FET
IMPATT	IMPact Avalanche Transit-Time
JFET (JUGFET)	junction FET
l.f.	low frequency
l.o.	local oscillator
MOST (MOSFET)	metal-oxide semiconductor FET
MRI	magnetic resonance imaging
NASA	National Aeronautics and Space Administration (USA)

n.f.b.	negative feedback
NMR	nuclear magnetic resonance
PIN	semiconductor device with p-i-n regions
p.l.l.	phase-locked loop
PPI	plan-position indicator
p.s.d.	phase sensitive detector
p.s.u.	power supply unit
r.f.	radio frequency
r.f.c.	radio frequency choke
r.m.s.	root mean square
SAW	surface-acoustic wave
SCR	silicon controlled rectifier
SNR	signal-to-noise ratio
TR	transmit-receive
TRAPATT	TRApped Plasma Avalanche Triggered Transit
t.r.f.	tuned radio frequency
TTL	transistor-transistor logic
u.h.f.	ultra high frequency
v.c.o.	voltage-controlled oscillator
VCR	voltage-controlled resistor
v.h.f.	very high frequency
VLSI	very large scale integration
1-D	one dimension/one-dimensional
2-D	two dimensions/two-dimensional
3-D	three dimensions/three-dimensional

Part 1
Electronics

This part of the book covers analogue electronics. It starts in Chapter 1 by describing the operation of the bipolar junction transistor and the field-effect transistor, and then (Chapters 2–4) developing various electronic building blocks and circuit configurations. In Chapters 5 and 6 negative and positive feedback topics are introduced. Perhaps illogically, operational amplifiers are applied (Chapter 7) before their internal design is discussed (Chapter 8), though this course is taken to clarify the specification and also to provide motivation before delving into design issues. Finally, Chapter 9 describes the design of stabilized power supplies, thereby rehearsing some of the principles of the earlier chapters.

1

Transistor Amplifying Devices

1.1 Introduction

Electronics is concerned with the design of amplifiers, oscillators, power supplies, radios, radar systems, telephones, tape recorders, computers and a host of other "black boxes" and systems. The internal workings of these machines depend on the existence and properties of a number of basic electronic components and devices. Early on in the development of the subject, available components included (for example) resistors, capacitors, metal rectifiers and thermionic valves. More recently, semiconductor devices such as transistors have largely replaced thermionic valves, and are themselves becoming incorporated into integrated circuits. This development has been important from the point of view of cost, speed, reliability, power consumption, convenience, and degree of miniaturization – to name but a few relevant factors. Nor are we nearing the end of this development, as electro-optical devices presage a new era of communication and computational hardware, while existing technologies are being pressed harder – and particularly towards greater degrees of miniaturization.

But what are the types of device that have allowed electronics to progress so far? Broadly, electronic components fall into two categories – passive and active devices. Passive devices include resistors, diodes, capacitors and inductors: these can be defined as components which have a simple relationship between current and voltage and which do not have a separate source of power that can be converted into another form (e.g. they are not supplied with a source of d.c. power which they can use to amplify the power of an a.c. signal). Active devices, on the other hand, include transistors, tunnel diodes and certain other devices which have more complex characteristics and are able to accept power in one form and to convert it to another form – amplification being a typical function performed by these devices.

It is tempting to assert that all two-terminal devices are passive, and that all devices with three or more terminals are active. In fact, this would be too simplistic, since attenuators can have many terminals, while tunnel diodes can be used as amplifiers or oscillators and are definitely active devices. On the other hand, the basic non-linear diode remains a passive device as it is only a type of two-terminal attenuator. We do not attempt any sort of rigorous definition or categorization, since the subject of electronics is full of surprises, and there are many devices which are beautifully simple yet which have highly interesting and useful properties (for example, variable-capacitance diodes, Zener diodes, constant current devices, unijunction transistors, and so on).

Thus it is clear that the subject has a richness borne of the many inventive people who have worked in it over the past 60 years or so. Yet resistors, capacitors, diodes and transistors must account for over 90% of the components that are used in electronic equipment (here we include the *individual* devices on integrated circuits). This means that any course on electronics must concentrate largely on what can be achieved with these four types of device. In what follows, we assume that the reader has already learnt about resistors, capacitors and diodes, and we start by examining the two main types of transistor.

1.2   The bipolar junction transistor

The bipolar junction transistor (BJT) is of vital importance in modern electronics. It can operate both as an amplifying element and as a switching element, depending on the type of circuit in which it is included. Here we concentrate on its amplifying properties.

The BJT is a three-terminal device, its three connections being known as the *emitter, base* and *collector*. We start by considering it as a pair of p-n junction diodes back to back. There are two ways in which this can be achieved, the result being either an npn or a pnp transistor (Figure 1.1). These have intrinsically identical properties, but differ merely in the polarities of the voltages and currents applied to them. Hence it will be sufficient if we concentrate initially on npn transistors.

In fact, it is *definitely* over-simplistic to regard the npn transistor as a trivial combination of two diodes, since this would make the transistor little more than an odd type of non-linear attenuator. Indeed, the transistor is constructed from two diodes in such close contact that they interact strongly

(a)

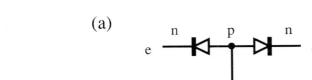

(b)

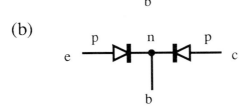

Fig. 1.1 npn and pnp transistors. This figure shows a rudimentary model of the bipolar junction transistor – that of two p-n junction diodes placed back to back. The two ways in which this can be achieved lead to (a) the npn and (b) the pnp transistor. e, emitter; b, base; c, collector.

to give a current gain rather than an attenuation. To achieve this, it is arranged that any electrons injected by the emitter into the p-type base region are almost certain to drift to the collector (see Appendix A for a summary of relevant properties of semiconductors). This will happen if (a) the base region is very thin, (b) the base region is only lightly doped, (c) the collector is made physically larger than either the base or the emitter, and (d) the voltage at the collector is made positive relative to the base. In addition, the base is made positive relative to the emitter, so that a substantial current is already flowing in the direction of the collector. This means that the emitter-base junction is forward-biased and the collector-base junction is reverse-biased (though there are occasions when, either deliberately or accidentally, a transistor is not biased in this way).

Under the conditions stated above, the majority of the electrons emitted into the base region eventually migrate to the collector (the remainder are captured by holes in the base region and result in a hole current at the base connection). Thus we can write:

$$I_c = \alpha I_e \tag{1.1}$$

where α is close to but necessarily slightly less than unity. This does not constitute a very useful current gain, but if we consider the collector current in relation to the base current the situation improves, since by Kirchhoff's current law:

$$I_e = I_c + I_b \tag{1.2}$$

We now deduce that:

$$I_c = \alpha(I_c + I_b) \tag{1.3}$$

Hence

$$I_c = \alpha I_b/(1 - \alpha) \tag{1.4}$$

i.e.

$$I_c = \beta I_b \tag{1.5}$$

where

$$\beta = \alpha/(1 - \alpha) \tag{1.6}$$

We also have the inverse relation:

$$\alpha = \beta/(\beta + 1) \tag{1.7}$$

Now α is in the range $0 < \alpha < 1$ and is normally close to unity; it is therefore clear that $\beta \gg 1$, typical figures being $\alpha \approx 0.995$ and $\beta \approx 200$. This shows that there is considerable amplifying capability between the base and collector terminals of the BJT. A similar situation applies for a pnp BJT, though here *holes* are injected by the emitter into the base region, and the majority of

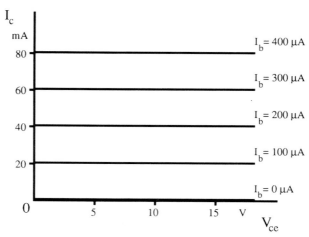

Fig. 1.2 Idealized collector characteristics of the BJT.

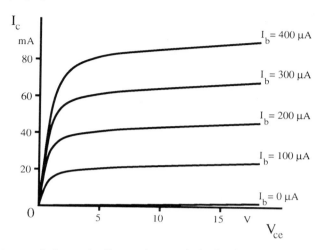

Fig. 1.3 More realistic set of collector characteristics for the BJT. These more realistic characteristics show the dependence of collector current I_c on collector voltage V_{ce}.

these appear at the collector – only a small proportion having been captured by electrons in the n-type base region.

Returning to equation (1.5), we observe that I_c depends only on I_b; there is apparently no dependence on V_{cb} or V_{ce}, and so we obtain the collector characteristics shown in Figure 1.2. It is important to realize that these characteristics are an idealization – as is equation (1.5), and equation (1.1) from which it is derived. In fact, the collector characteristics *do* exhibit a slight dependence on V_{ce}, and in addition β varies slightly with I_c. The general forms of these dependences are indicated in Figures 1.3 and 1.4 respectively. However, in most of what follows the collector characteristics can be modelled accurately enough as straight lines, except at low values of V_{ce}, and the variation of β can be ignored.

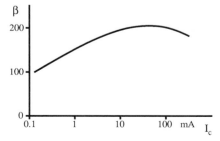

Fig. 1.4 Variation of β current gain for a BJT.

The next problem is how to turn the large (β) current gain into a useful voltage gain. Consider the simple circuit shown in Figure 1.5. A positive input voltage V_1 will clearly produce a substantial base current, and the collector current will be much amplified and will generate a large output voltage across R_c. However, we cannot yet calculate the voltage gain A_V, since we do not know the input impedance of the transistor. In addition, we can expect that the input impedance will be highly non-linear, since the emitter-base diode closely obeys the well-known exponential law for a junction diode:

$$I = I_0[e^{eV/kT} - 1] \tag{1.8}$$

We shall obtain a more accurate model of the BJT below. Meanwhile, we calculate the voltage gain for an amplifier which includes a sizeable resistor R_1 in series with the base:

$$I_b = V_1/R_1 \tag{1.9}$$

$$\therefore \ V_{ce} = V_{cc} - I_c R_c = V_{cc} - \beta I_b R_c = V_{cc} - \beta V_1 R_c/R_1 \tag{1.10}$$

Hence

$$A_V = \Delta V_{ce}/\Delta V_1 = -\beta R_c/R_1 \tag{1.11}$$

and the large value of β means that quite large voltage gains should be achievable in practical situations (but note that we have so far taken R_1 to be much larger than the inherent input resistance of the BJT).

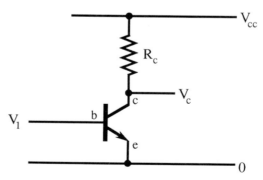

Fig. 1.5 Simple transistor "common emitter" (CE) amplifier. This circuit is analysed in detail in the text. However, it should be noted that it is normally used in modified form, e.g. with a resistor R_1 in series with the base of the transistor, in order to limit the base current. e, b, c indicate respectively the emitter, base and collector of the BJT: note that the emitter of a BJT is always marked with an arrow (for a pnp BJT the direction of the arrow would be reversed).

1.3   Obtaining a more detailed model of transistor operation

It should be noted that equation (1.5) is only valid when the transistor is operating in its linear region. In that case we must have I_b, $I_c \geqslant 0$, and $V_c > V_b > V_e$. Clearly, outside this range equation (1.10) also becomes invalid, and in practice this means that V_{ce} cannot swing above V_{cc} or below the saturation voltage of about 0.1 V (see Figure 1.6). (Formally, a transistor is said to be *cut-off* when $I_c = 0$, and *saturated* when $V_b > V_e$ *and* $V_b > V_c$, so that electrons are injected into the base region from both the emitter and the collector. At that point a considerable amount of stored charge is built up in the base region, and there is some difficulty in removing it rapidly.)

A_V may be defined as the large-signal voltage gain, and this circuit is generally used only for amplifying large signals or voltage pulses. When small signals are to be amplified, it is normally important for the amplifier to be biased into its linear region and for the small input signals to be fed to it via a capacitor. There is also another class of amplifier in which small d.c. signals are to be amplified: in that case, the transistor is again biased to the linear part of its range, but capacitors are not employed. In what follows we shall concentrate first on a.c. amplifier design. Equation (1.11) makes it clear that such an amplifier will have maximum gain if R_1 is reduced towards zero. Hence it is important to know the input impedance of the BJT itself. We now attempt to determine its value in the case of small signals.

Applying the diode exponential law to the emitter-base junction diode, we have:

$$I_b = I_e = I_{eb0}\left[\exp\left(\frac{eV_{be}}{kT}\right) - 1 \right] \qquad (1.12)$$

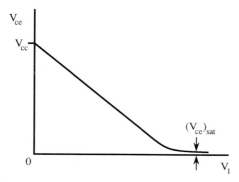

Fig. 1.6   Forward voltage transfer characteristic for a BJT.

an equation that is clearly only valid when $I_c = 0$. When $I_c > 0$, $I_b \ll I_e$, and equation (1.12) describes I_e but not I_b (ultimately, this is because the doping of the transistor base and emitter regions is such that most of the emitter-base diode current is conducted by carriers which start from the emitter). We now deduce that:

$$\frac{dI_e}{dV_{be}} = I_{eb0} \times \left(\frac{e}{kT}\right) \times \exp\left(\frac{eV_{be}}{kT}\right) \approx \left(\frac{e}{kT}\right)I_e \tag{1.13}$$

Thus the *basic* a.c. input impedance of the transistor is:

$$r_e = \frac{kT}{eI_e} \approx \frac{1}{40I_e} \tag{1.14}$$

leading to a value of about $1\,\Omega$ for $I_e \approx 25\,\text{mA}$.

However, r_e is not the only factor contributing to the input impedance of the BJT, and the intrinsic (ohmic) resistance of the transistor material is also important. In particular, the resistance r_b of the base region has to be considered. Finally, if we are to have a realistic model of the transistor as a whole, the "drift resistance" r_d, which expresses the fact that changes in collector voltage slightly modify the width of the base region and hence affect the number of electrons reaching the collector, must be taken into account. This gives us the T-equivalent circuit of a BJT shown in Figure 1.7. Note that for a BC107 silicon transistor operating under fairly standard conditions, viz.

$$V_{ce} = 10\,\text{V}$$

$$V_{be} = 0.6\,\text{V}$$

$$I_c = 25\,\text{mA}$$

$$I_b = 100\,\mu\text{A}$$

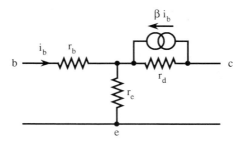

Fig. 1.7   *T*-equivalent circuit of the BJT.

the four T-parameters have values:

$$r_e \approx 1\,\Omega$$

$$r_b \approx 750\,\Omega$$

$$r_d \approx 3.3\,k\Omega$$

$$\beta \approx 250$$

It should be remembered that these parameters vary considerably with the operating conditions, and somewhat unpredictably as different transistors even of the same type are chosen (except in the case of matched transistors on the same substrate). Thus for the three types of BC107:

A: $\beta \approx 125-260$

B: $\beta \approx 240-500$

C: $\beta \approx 450-900$

Although the T-equivalent circuit gives a realistic physical model of the BJT, it does not immediately provide a practical model that clarifies quantities such as the input impedance. To achieve this, a number of other models are available whose parameters are related to those of the T-equivalent circuit. In particular, we shall consider the h-parameter and y-parameter models. All such models have four intrinsic parameters, since there are four currents and voltages to be related – i_b, i_c, v_{be}, v_{ce} – two of which can be chosen to be expressed in terms of the other two. In the h-parameter case, i_c and v_{be} are expressed in terms of i_b and v_{ce}, whereas in the y-parameter case i_b and i_c are expressed (rather more systematically!) in terms of v_{be} and v_{ce}. First we consider the h-parameter model.

1.4 h-Parameters in terms of T-parameters

Figure 1.8 shows a T-parameter and an h-parameter model of the BJT which are to be considered equivalent. For the T-parameter model we have:

$$v_{be} - v = i_b r_b \tag{1.15}$$

$$v_{ce} - v = (i_c - \beta i_b) r_d \tag{1.16}$$

$$v = (i_b + i_c) r_e \tag{1.17}$$

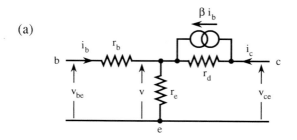

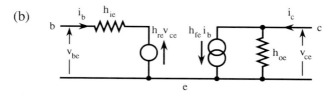

Fig. 1.8 Comparison of (a) T-parameter and (b) h-parameter models of the BJT.

and for the h-parameter model we have:

$$v_{be} = h_{ie}i_b + h_{re}v_{ce} \tag{1.18}$$

$$i_c = h_{fe}i_b + h_{oe}v_{ce} \tag{1.19}$$

where the suffices i, o, f, r, and e on the h-parameters denote respectively *input*, *output*, *forward* (i.e. from input to output), *reverse* (i.e. from output back to input), and *common emitter* connection (i.e. the emitter provides the common or ground line between the input and the output). Eliminating v between the T-parameter equations (1.15–1.17) gives:

$$v_{be} = i_b(r_e + r_b) + i_c r_e \tag{1.20}$$

$$v_{ce} = i_b(r_e - \beta r_d) + i_c(r_e + r_d) \tag{1.21}$$

We now need to re-cast these equations in h-equation form, i.e. express v_{be} and i_c in terms of i_b and v_{ce}. To achieve this we eliminate i_c and v_{be} in turn, obtaining respectively:

$$v_{be} = i_b(r_e + r_b) + \left(\frac{r_e}{r_e + r_d}\right)[v_{ce} - i_b(r_e - \beta r_d)] \tag{1.22}$$

$$i_c = i_b\left(\frac{\beta r_d - r_e}{r_e + r_d}\right) + \frac{v_{ce}}{(r_e + r_d)} \tag{1.23}$$

Equating coefficients with the h-parameter equations (1.18–1.19) now gives:

$$h_{ie} = (r_e + r_b) + \left(\frac{r_e}{r_e + r_d}\right)(\beta r_d - r_e) \approx r_b + (\beta + 1)r_e \qquad (1.24)$$

$$h_{re} = \left(\frac{r_e}{r_e + r_d}\right) \approx \frac{r_e}{r_d} \qquad (1.25)$$

$$h_{fe} = \left(\frac{\beta r_d - r_e}{r_e + r_d}\right) \approx \beta \qquad (1.26)$$

$$h_{oe} = \left(\frac{1}{r_e + r_d}\right) \approx \frac{1}{r_d} \qquad (1.27)$$

These approximations assume only that $r_d \gg r_e$, which is always valid – as in the case of the BC107 npn transistor whose T-parameters were listed in Section 1.3. Typical h-parameter values corresponding to this set of T-parameters are:

$$h_{ie} = 1 \text{ k}\Omega$$

$$h_{re} = 3 \times 10^{-4}$$

$$h_{fe} = 250$$

$$h_{oe} = 300 \text{ }\mu\text{S}$$

Note the mixed set of dimensions of these parameters: this leads to the term *hybrid* for this set of parameters.

1.5 y-Parameters in terms of h-parameters

Figure 1.9 shows the y-parameter model of the BJT. In this case the appropriate equations are:

$$i_b = y_{ie}v_{be} + y_{re}v_{ce} \qquad (1.28)$$

$$i_c = y_{fe}v_{be} + y_{oe}v_{ce} \qquad (1.29)$$

Note that all four y-parameters have the dimensions of conductance – a situation which contrasts markedly with the case of the h-parameters.

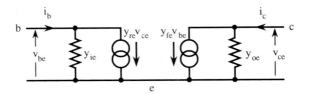

Fig. 1.9 *y*-parameter model of the BJT.

Some rather simple analysis, along the lines used above for finding the *h*-parameters in terms of the *T*-parameters, gives the following formulae for the *y*-parameters in terms of the *h*-parameters:

$$y_{ie} = \frac{1}{h_{ie}} \tag{1.30}$$

$$y_{re} = -\frac{h_{re}}{h_{ie}} \tag{1.31}$$

$$y_{fe} = \frac{h_{fe}}{h_{ie}} \tag{1.32}$$

$$y_{oe} = h_{oe} - \frac{h_{re}h_{fe}}{h_{ie}} \tag{1.33}$$

Note that h_{re} and y_{re} can be taken as negligible compared with the other parameters. (Strictly, such statements can only be made rigorously in the *y*-parameter case, since all *y*-parameters are expressed in the same units – see below.) Assuming this, inspection of the two circuit models immediately gives:

$$y_{ie} = \frac{1}{h_{ie}} \tag{1.34}$$

$$y_{re} \approx 0; \qquad h_{re} \approx 0 \tag{1.35}$$

$$y_{fe} = \frac{h_{fe}}{h_{ie}} \tag{1.36}$$

$$y_{oe} \approx h_{oe} \tag{1.37}$$

Typical values for the y-parameters (under the standard operating conditions quoted in Section 1.3) are:

$$y_{ie} = 1 \, mS$$

$$y_{re} = -0.3 \, \mu S$$

$$y_{fe} = 0.25 \, S$$

$$y_{oe} = 225 \, \mu S$$

As pointed out earlier, the relative values of the y-parameters can be compared quite rigorously, since they all have the same dimensions; in fact we see that:

$$y_{fe} \gg y_{ie} \gg y_{oe} \gg |y_{re}| \tag{1.38}$$

In addition, as we shall see in Section 1.8, the y-parameters, and in particular y_{fe}, are very useful in forming a means by which to compare the BJT and the FET – and also other amplifying devices such as thermionic valves.

Finally, we notice that if r_b is small:

$$y_{fe} = \frac{h_{fe}}{h_{ie}} \approx \frac{\beta}{(\beta+1)r_e} \approx \frac{e}{kT} \times \frac{\beta I_e}{(\beta+1)} = \frac{e}{kT} \times I_c = 40 I_c \tag{1.39}$$

For the "typical" value of collector current of 25 mA quoted above, this gives a value of $y_{fe} = 1 \, S$ – though at (for example) 1 mA it is reduced to just 40 mS.

1.6 A basic amplifier circuit

Now that we have a more comprehensive model of the BJT, it is possible to analyse a highly useful amplifier circuit. The circuit we shall take is in fact slightly more general than the one considered earlier, and includes a resistor R_e in series with the emitter (Figure 1.10). In this case we have:

$$i_b = (v_1 - v_e)/h_{ie} \tag{1.40}$$

$$i_c = \beta i_b \tag{1.41}$$

$$i_e = i_c + i_b = v_e/R_e \tag{1.42}$$

$$v_c = -i_c R_c \tag{1.43}$$

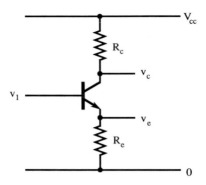

Fig. 1.10 A basic 1-transistor amplifier. This circuit has some importance, since it is a generalized form of both the common emitter amplifier (for which $R_e=0$) and the emitter follower circuit (for which $R_c=0$). This makes it worth analysing its operation in some detail (see text).

where equation (1.43) arises on differentiating the large-signal equation:

$$V_c = V_{cc} - I_c R_c \qquad (1.44)$$

First, note that we can easily find v_c in terms of v_e, since equations (1.41–1.43) show that:

$$v_c = -\beta i_b R_c = -\left(\frac{\beta}{\beta+1}\right) i_e R_c = -v_e \left(\frac{\beta}{\beta+1}\right) \frac{R_c}{R_e} \qquad (1.45)$$

Since normally $\beta \gg 1$, we find:

$$\frac{v_c}{v_e} \approx -\frac{R_c}{R_e} \qquad (1.46)$$

An interesting situation arises if $R_e = R_c$. In that case:

$$v_c \approx -v_e \qquad (1.47)$$

and the circuit can be used as a phase-splitter, with two outputs of almost equal amplitude and opposite phase. Such circuits are useful for feeding high power output (push-pull) stages, and also for driving variable phase networks, as we shall see later.

Now that we know v_c in terms of v_e, it remains to find v_e in terms of v_1. Eliminating i_b and i_c between equations (1.40–1.42) gives:

$$v_e = (\beta + 1)i_b R_e = (\beta + 1)R_e(v_1 - v_e)/h_{ie} \tag{1.48}$$

$$\therefore \quad v_1 = v_e[1 + h_{ie}/R_e(\beta + 1)] \tag{1.49}$$

Hence

$$\frac{v_e}{v_1} = \frac{1}{1 + h_{ie}/R_e(\beta + 1)} \tag{1.50}$$

and if $\beta \gg 1$, and R_e is not too small, we find that

$$v_e \approx v_1 \tag{1.51}$$

which is also the standard result for an emitter follower (for which $R_c = 0$, as in Figure 1.11): we shall return to this case below. Combining this with our previous result, we find:

$$\frac{v_c}{v_1} \approx -\frac{R_c}{R_e} \tag{1.52}$$

So far we have ignored the possibility that R_e might be small. In fact if $R_e = 0$, we find:

$$v_e = 0 \tag{1.53}$$

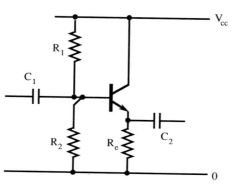

Fig. 1.11 Emitter follower. This figure shows the a.c. version of the emitter follower. The d.c. version is remarkably simple, and just includes the transistor and emitter resistor R_e.

and

$$\frac{v_c}{v_1} = -\frac{\beta R_c}{h_{ie}} \qquad (1.54)$$

It is interesting that this last equation can be derived much more simply using y-parameters. We first calculate i_c using the formula:

$$i_c = y_{fe} v_1 \qquad (1.55)$$

Then we can deduce v_c:

$$v_c = -i_c R_c = -y_{fe} R_c v_1 \qquad (1.56)$$

Next, we recall equation (1.32) giving y_{fe} in terms of h-parameters, and then note (equation 1.26) that h_{fe} is approximately equal to β. This immediately gives equation (1.54) above.

Equation (1.54) applies for the circuit we might wish to use when high gain is required. Inserting the typical figures mentioned above, we find that the small-signal voltage gain $a_v \approx -500$ when $R_c = 2\,k\Omega$ (though a somewhat smaller value of a_v would be obtained if full account were taken of the output admittance h_{oe} of the transistor). However, the case where $R_e > 0$ is of some value, since when $R_e \gg r_e$ the gain will be approximately independent of the transistor parameters. We shall see later that this is an instance of the application of negative feedback, which is able to stabilize gain and also to cut down distortions due to the non-linearity of the transistor emitter-base junction. In addition, this circuit has another interesting function: it is able to stabilize the d.c. operating point when the circuit is used as a high gain a.c. amplifier. In this case the complete circuit employs a number of capacitors, as shown in Figure 1.12.

A full analysis of the circuit of Figure 1.12 shows that significant loss of a.c. gain will occur if the impedance of the emitter capacitor C_2 is too high. A priori, we might have thought that Z_{C_2} must merely be made much less than R_e, but in fact the situation is more stringent than this, and Z_{C_2} must be much less than the parallel combination of R_e and r_e; this is because r_e is (at worst) a direct path to earth from the emitter connection (see the T-equivalent circuit in Figure 1.7), and is effectively in parallel with $C_2{}^*$. Thus:

$$Z_{C_2} \ll R_e \| r_e \qquad (1.57)$$

---

* Another way of looking at the problem, which gives a more exact answer, is that C_2 is effective at those frequencies for which $Z_{C_2} \ll Z_{out}$ for the emitter follower, the value of Z_{out} being given by equation (1.65).

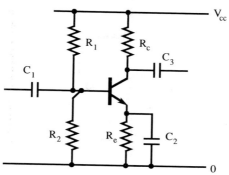

Fig. 1.12 Practical form of the 1-transistor common emitter amplifier. In this commonly used a.c. amplifier circuit, the d.c. operating point is stabilized by negative feedback (see text). Typical values of the resistors are: $R_1 = 3.9\,\mathrm{k\Omega}$, $R_2 = 1\,\mathrm{k\Omega}$, $R_e = 100\,\Omega$ and $R_c = 270\,\Omega$. If $V_{cc} = 12\,\mathrm{V}$, then $V_b \approx 2.5\,\mathrm{V}$, $V_e \approx V_b - 0.6 \approx 2\,\mathrm{V}$, and $V_c \approx 12 - 2 \times 270/100 \approx 6.5\,\mathrm{V}$, giving nearly the maximum available a.c. swing of $\pm 5\,\mathrm{V}$. With resistors normally only being within 5% of their marked values, high accuracy in such calculations is unnecessary.

However, the input capacitor C_1 only has to have low impedance relative to h_{ie}:

$$Z_{C_1} \ll h_{ie} \qquad (1.58)$$

Finally, the output capacitor C_3 has to have low impedance relative to the parallel combination of R_c and r_d:

$$Z_{C_3} \ll R_c \| r_d \qquad (1.59)$$

All three conditions listed above are frequency-dependent: this means that if f_{min} is the lowest frequency at which the circuit is to operate, we must have:

$$C_1 \gg 1/2\pi f_{min} h_{ie} \qquad (1.60)$$

$$C_2 \gg 1/2\pi f_{min}(R_e \| r_e) \qquad (1.61)$$

$$C_3 \gg 1/2\pi f_{min}(R_c \| r_d) \qquad (1.62)$$

1.7 Input and output impedance of the emitter follower

We now return to the emitter follower, which is the special case of our general circuit when $R_c = 0$ (Figure 1.11). Above we proved that in this case $v_e \approx v_1$: at first sight this appears not to be a useful result, but in fact it is exceptionally useful, because of the low output impedance of the circuit. This property results again from negative feedback, which arises since if the load across R_e takes excessive current, the voltage across R_e will fall, v_{be} will rise, i_b will also rise; hence i_e will increase, and v_e will tend to return to its proper value close to v_1. For similar reasons the circuit turns out to have a high input impedance.

To find the output impedance Z_{out}, we need to express the output voltage in terms of an ideal voltage generator in series with Z_{out}. We can achieve this by re-expressing equation (1.50) in the form:

$$\frac{v_e}{v_1} = \frac{R_e}{R_e + h_{ie}/(\beta+1)} = \frac{R_e}{R_e + Z_{out}} \tag{1.63}$$

where

$$Z_{out} = h_{ie}/(\beta+1) \approx r_e + r_b/(\beta+1) \tag{1.64}$$

However, this gives only part of the answer. The fact that r_b appears here implies that if the input circuit is fed from a circuit of source impedance R_s, this will be added to r_b. Similarly, if the emitter follower load is not R_e but R_L, then the impedance driving R_L will be reduced by the parallel effect of R_e. Taking account of both of these factors now gives a more complete formula for output impedance:

$$Z_{out} = [(h_{ie} + R_s)/(\beta+1)] \| R_e = [r_e + (r_b + R_s)/(\beta+1)] \| R_e \tag{1.65}$$

The input impedance is obtained by determining how i_b varies with v_1. From equations (1.40) and (1.63) we get:

$$i_b = \frac{v_1(1 - v_e/v_1)}{h_{ie}} = v_1 \frac{1}{R_e(\beta+1) + h_{ie}} \tag{1.66}$$

$$\therefore \ Z_{in} = R_e(\beta+1) + h_{ie} \approx (R_e + r_e)(\beta+1) + r_b \tag{1.67}$$

Again, a modified form of this equation will be appropriate if the true load is a resistor R_L placed across R_e:

$$Z_{in} = (R_e \| R_L)(\beta+1) + h_{ie} = (R_e \| R_L + r_e)(\beta+1) + r_b \tag{1.68}$$

However, the most important feature of the final equations for input and output impedance is that there is a buffering factor of $\beta+1$, so that the *output* load resistance is magnified by a factor $\beta+1$ when seen from the *input* of

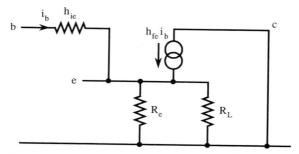

Fig. 1.13 Equivalent circuit for determining input impedance. This figure shows the equivalent circuit of the emitter follower of Figure 1.11, required for calculating the *input* impedance.

the circuit, and the *input* source resistance is reduced by a factor $\beta+1$ when seen from the *output*.

Some further notes on the above equations are in order:

1. It is rather satisfying that the formulae for Z_{in} and Z_{out} can be understood (and indeed, derived somewhat more simply) by analysing the equivalent circuits in Figures 1.13, 1.14: for example, Figure 1.13 shows that the voltage drop that arises across the input from the current i_b is $h_{ie}i_b + (h_{fe}+1)i_b(R_e\|R_L)$, and this must be equal to $Z_{in}i_b$.

2. The above equations for Z_{in} and Z_{out} are idealized and ignore the effects of the bias resistors R_1, R_2. However, it is easy to include the effects of these resistors in the above formulae – e.g. by suitably modifying the value of R_s when calculating Z_{out}.

3. When h_{ie} is replaced by $r_b + r_e(\beta+1)$, the resistances r_b, r_e appear in intuitively obvious places in the equations for Z_{in} and Z_{out}.

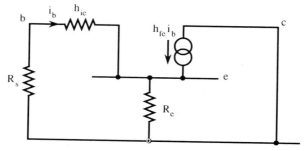

Fig. 1.14 Equivalent circuit for determining output impedance. This figure shows the equivalent circuit of the emitter follower of Figure 1.11, required for calculating the *output* impedance.

Finally, note that the large-signal solution for an emitter follower has the approximate form:

$$V_e = V_1 - V_{be} \approx V_1 - 0.6 \qquad (1.69)$$

where the 0.6 V offset is due to the effective switch-on voltage of the emitter-base diode.

1.8   The field-effect transistor

The field-effect transistor (FET) is another useful device, which has a number of functions varying from amplification and switching to square-law generation. It is again a three-terminal device, the three connections being known as the *source, gate* and *drain*. In this case there is a *channel* of n-type semiconductor joining the source and drain connections, and the gate connection is joined to a p-type region which forms a p-n junction diode with the channel. Normally, the gate junction is reverse-biased, and the gate voltage influences the flow of electrons between source and drain. The electrons are majority carriers in the n-type channel, and this contrasts with the BJT, where the carriers from emitter to collector are minority carriers in the base region. Furthermore, in the FET the majority carriers determine the properties of the device and therefore it is sometimes called a *unipolar* device: this again contrasts with the BJT where both the minority and majority carriers are important in determining its operation – hence the term *bipolar*. These points are enlarged upon in Appendix A.

Naturally, there are two possible polarities of FET, these being classed as n- or p-channel FETs. In addition, there are *insulated gate* FETs (IGFETs)*, which have much higher input impedance: to contrast with these, the type of FET described above is called a junction FET (JFET, or sometimes JUGFET).

It is somewhat more difficult to model the characteristics of the FET than those of the BJT. However, it is fairly obvious that the drain current will depend largely on the gate voltage, i.e. $I_d = I_d(V_{gs})$. In fact, it turns out that, to a good approximation for a junction FET:

$$I_d \approx I_{dss}\left(1 + \frac{V_{gs}}{V_p}\right)^2 \qquad (1.70)$$

---

* Also known as *metal-oxide semiconductor* FETs (MOSFETs or MOSTs), because of their construction.

where I_{dss} is the drain current for $V_{gs} = 0$; V_p is called the "pinch-off" voltage, and is the *magnitude* of the gate voltage at which the drain current is (apart from a small leakage current) reduced to zero: the actual value of V_{gs} at which this happens is negative and equals $-V_p$*. At that point the majority carriers in the channel are repelled by the voltage on the gate to such an extent that the width of the conducting region in the channel effectively drops to zero. Typical values of V_p and I_{dss} are 5 V and 10 mA respectively. Note that the value of the index in equation (1.70) can in practice take values between ~ 1.9 and 2.1.

As in the case of the BJT, the drain characteristics are not exactly parallel to the V_{ds} axis, but at least they can be approximated by parallel straight lines in the "saturation" region where V_{ds} is not too small (Figure 1.15). On the other hand, in the "ohmic" region, where V_{ds} is small, the characteristics approximate to straight lines passing through the origin, and this opens up the possibility of the FET being used as a voltage-controlled resistor or VCR. It is of interest for such applications that the ohmic region extends through the origin. To calculate the resistance of the channel near the origin we start

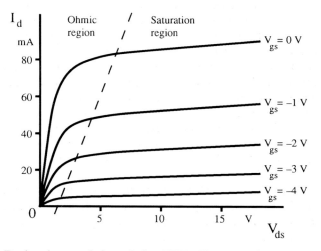

Fig. 1.15 Drain characteristics of the FET. These continuous curves can be considered to consist of two regions – the "ohmic" region at low values of V_{ds} and the "saturation" region at higher values of V_{ds}. In the latter region, the curves are almost independent of V_{ds}.

---

* Note that some authors take the alternate definition, with V_p being negative.

with a formula giving the drain current over the ohmic region:

$$I_d \approx \frac{2I_{dss}V_{ds}}{V_p}\left(1 + \frac{V_{gs}}{V_p} - \frac{V_{ds}}{2V_p}\right) \tag{1.71}$$

Hence

$$R_{ds} \approx R_{ds0}\left(1 + \frac{V_{gs}}{V_p} - \frac{V_{ds}}{2V_p}\right)^{-1} \approx R_{ds0}\left(1 + \frac{V_{gs}}{V_p}\right)^{-1} \tag{1.72}$$

where

$$R_{ds0} = V_p/2I_{dss} \tag{1.73}$$

We now move on to consider the small-signal amplifying characteristics of the FET. The y-parameters are of particular importance for this purpose. We have:

$$i_g = y_{is}v_{gs} + y_{rs}v_{ds} \tag{1.74}$$

$$i_d = y_{fs}v_{gs} + y_{os}v_{ds} \tag{1.75}$$

For definiteness, we quote typical y-parameters for an n-channel junction FET similar to a 2N3819 under the operating conditions $V_{ds} = 10\,\text{V}$, $V_{gs} = -2\,\text{V}$ and $I_d = 4\,\text{mA}$:

$$y_{is} \approx 10^{-10}\,\text{S}$$

$$y_{rs} \approx 0$$

$$y_{fs} = 2.5\,\text{mS}$$

$$y_{os} = 30\,\mu\text{S}$$

The y-parameters, and in particular y_{fs}, are very useful in forming a means by which to compare the BJT and the FET – and also other amplifying devices such as thermionic valves. It will be clear that the FET parameters show significant differences from those for the BJT given earlier. Note particularly that y_{fs} is typically 20–100 times smaller than y_{fe} for the BJT (though both of these y_f parameters vary markedly with the relevant output current). In addition:

$$y_{fs} \gg y_{os} \gg y_{is} \gg y_{rs} \qquad (1.76)$$

Normally we can clearly ignore i_g and employ just the equation for i_d. Next, recalling the approximation for I_d given by equation (1.70), we deduce:

$$y_{fs} = \frac{dI_d}{dV_{gs}} \approx \frac{2I_{dss}}{V_p}\left(1 + \frac{V_{gs}}{V_p}\right) = \frac{2(I_d I_{dss})^{1/2}}{V_p} \qquad (1.77)$$

Taking $V_p = 5$ V and $I_{dss} = 10$ mA, as quoted earlier, we deduce that y_{fs} will typically vary from 0 at $I_d = 0$ to ~ 4 mS at $I_d = I_{dss}$.

Finally, note that when this equation is extrapolated (see Figure 1.15) to $V_{gs} = 0$:

$$y_{fs} \approx 2I_{dss}/V_p \qquad (1.78)$$

By some curious circumstance, this value of y_{fs} is equal to the reciprocal of R_{ds0} (equation 1.73), and in addition the slightly more general relation

$$y_{fs} \approx 1/R_{ds} \qquad (1.79)$$

is also valid for other values of V_{gs} (see equations 1.72 and 1.77) – i.e. the *conductance* in the ohmic region is equal to the *forward conductance*, at the same value of V_{gs}, in the saturation region. (Note that the forward conductance is also widely known as the *mutual conductance* which is often denoted by g_m.)

1.9 The FET source follower

In this section we briefly examine the FET source follower circuit and compare its properties with those of the BJT emitter follower. Figure 1.16 shows the source follower. Since y_{rs} can be ignored and y_{is} is very small, we can proceed with the following equations:

$$i_d = y_{fs}v_{gs} + y_{os}v_{ds} \qquad (1.80)$$

$$i_d R_s = v_s \qquad (1.81)$$

$$v_1 = v_s + v_{gs} \qquad (1.82)$$

$$v_s + v_{ds} = 0 \qquad (1.83)$$

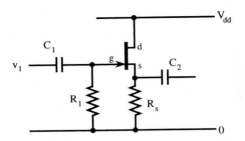

Fig. 1.16 FET source follower circuit. The main difference between the FET source follower and the BJT emitter follower is the single bias resistor R_1 used with the FET. R_1 is typically $\sim 1\,\mathrm{M\Omega}$. s, g, d indicate respectively the source, gate and drain of the FET: note that the gate of the FET is marked with an arrow (for a p-channel FET the direction of the arrow would be reversed).

Eliminating v_{gs}, v_{ds} and i_{d}, we find*:

$$v_{\mathrm{s}}/R_{\mathrm{s}} = y_{\mathrm{fs}}(v_1 - v_{\mathrm{s}}) - y_{\mathrm{os}}v_{\mathrm{s}} \tag{1.84}$$

$$\therefore \quad (y_{\mathrm{fs}} + y_{\mathrm{os}} + 1/R_{\mathrm{s}})v_{\mathrm{s}} = y_{\mathrm{fs}}v_1 \tag{1.85}$$

$$\therefore \quad \frac{v_{\mathrm{s}}}{v_1} = \frac{y_{\mathrm{fs}}}{y_{\mathrm{fs}} + y_{\mathrm{os}} + 1/R_{\mathrm{s}}} \approx \frac{y_{\mathrm{fs}}}{y_{\mathrm{fs}} + 1/R_{\mathrm{s}}} \tag{1.86}$$

and for $y_{\mathrm{fs}} \gg 1/R_{\mathrm{s}}$ we find that the voltage gain is close to unity. However, for common values of y_{fs} and R_{s} such as $2.5\,\mathrm{mS}$ and $1\,\mathrm{k\Omega}$, the voltage gain is $2.5/(2.5 + 1) \approx 0.7$, and it is rarely much greater than 0.9.

Now consider the output impedance. Re-arranging equation (1.86), we find:

$$\frac{v_{\mathrm{s}}}{v_1} = \frac{R_{\mathrm{s}}}{R_{\mathrm{s}}(y_{\mathrm{fs}} + y_{\mathrm{os}})/y_{\mathrm{fs}} + 1/y_{\mathrm{fs}}} \approx \frac{R_{\mathrm{s}}}{R_{\mathrm{s}} + 1/y_{\mathrm{fs}}} \tag{1.87}$$

Hence, if R_{s} is considered to be the load, $Z_{\mathrm{out}} \approx 1/y_{\mathrm{fs}}$, but if there is a separate load, R_{s} will be in parallel with it, and hence the output impedance becomes:

$$Z_{\mathrm{out}} \approx R_{\mathrm{s}} \| 1/y_{\mathrm{fs}} \tag{1.88}$$

---

*Notice that the final equation (1.86) for the gain of the source follower can essentially be derived direct from the corresponding equation (1.63) for the BJT by re-expressing the h-parameters in terms of y-parameters and substituting the FET y-parameters.

The typical values of R_s and y_{fs} given above therefore lead to an output impedance of about $300\,\Omega$, which is much larger than would be expected for an emitter follower. However, if an emitter follower is fed from a generator of output impedance greater than $\beta \times 300\,\Omega$, this will dominate the emitter follower output impedance and the latter will be higher than for the FET source follower circuit. Thus there are some occasions when a source follower will have a better (i.e. lower) output impedance than an emitter follower, as well as a better (i.e. higher) input impedance (the latter will basically be in the region of $10^{10}\,\Omega$ for a source follower at low frequencies, though the gate bias resistor will lower this substantially).

1.10 Other FET circuits

It will not be possible to cover all aspects of FET circuit design in this book. For the most part we will have to content ourselves in the knowledge that most BJT circuits have FET counterparts with not too dissimilar characteristics. However, one or two observations are in order.

First, the FET should be regarded as a voltage-operated rather than a current-operated device. This is because its operation is determined largely by its voltage amplifying properties which are characterized by the mutual conductance y_{fs}: by comparison the BJT is intrinsically* a current amplifier with current gain h_{fe}. These statements are underlined by the high input impedance of the FET and the rather low input impedance of the BJT. Clearly, the required input impedance can be an important factor in deciding which of these types of device is to be used in a given application.

Second, the FET is normally operated with its input (gate) voltage lower than its common (source) voltage, whereas for the BJT the opposite is true. This makes self-biasing rather more straightforward for the FET, in that the gate is normally connected to ground via a single large resistance of $1\,M\Omega$ or more – compare Figures 1.12 and 1.17. The result is that FET circuits often appear superficially different from their BJT counterparts.

Third, the FET has some specific properties that make it especially suitable for certain types of application. Apart from its high input impedance, and its capability for use as a VCR – properties that have already been referred

---

*However, modern thinking is moving away from this view. The reason for this is that the current gain β varies significantly with I_c, whereas y_{fe} is nearly constant over many decades, indicating that the exponential law for the emitter-base junction is the dominant factor controlling device performance.

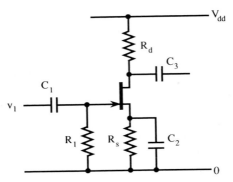

Fig. 1.17 FET common source amplifier. This circuit should be compared with the common emitter amplifier of Figure 1.12. Note the single bias resistor R_1 used with the FET circuit. R_1 is typically $\sim 1\,\mathrm{M\Omega}$.

to – its square-law drain current variation makes it suitable for certain analogue processing functions (see Section 2.8). In addition, its simple bias circuit permits it to be used as an integrated constant current device (see Section 3.1). Finally, its noise performance is in some cases superior to that of the BJT, though the full picture is quite complex and great care has to be taken in the choice of a suitable low-noise amplifying device (see Chapter 11).

1.11 Summary

This chapter has covered two important amplifying devices – the BJT and the FET. A variety of models of the BJT exists, the T-parameter model (Section 1.3) being quite closely related to its physical parameters, and the h-parameter model (Section 1.4) being widely used as a phenomenological model. These two models, and the y-parameter model (which is of particular value when describing the FET), have been related mathematically in Sections 1.4 and 1.5, in order to permit the detailed analysis of BJT circuits.

Section 1.6 has discussed in some detail a basic amplifier circuit using the BJT. This circuit is a valuable one to consider, since it yields the common emitter amplifier and the emitter follower as special cases. Section 1.6 also covered the a.c. version of the common emitter amplifier, while Section 1.7 examined the emitter follower and its input and output impedances – the high and low values of these respective quantities being the main virtues of this circuit. Sections 1.8 and 1.9 went on to examine the FET and circuits using it, including the common source amplifier and the source follower: it also covered properties of the FET which lead on to its use in voltage-controlled resistors, constant current sources, square-law function generators and low-noise amplifiers.

For reasons of space, this chapter has concentrated mainly on the BJT. However, this treatment is not intended to leave the impression that practitioners of electronics should ignore the FET. On the contrary, in electronics as in other engineering subjects, there are many tools in the toolbox, and the right one should always be selected to fulfil the task in hand – there being many cases where the FET fulfils the requirements of a particular practical application (see the list of uses of the FET given at the end of the last paragraph).

1.12 Bibliography

The present volume is intended to cover the basics of analogue electronics, but does not have space to discuss the properties of semiconductors, or elementary electricity, in any depth (though Appendix A provides *some* background on properties of semiconductors): for further information on these two topics the reader is referred to Kittel (1966), Sze (1981), and Calvert and McCausland (1978). Watson (1989) provides a more advanced discussion on transistor amplifying devices and their circuit models (including the hybrid-π equivalent circuit used to model transistors at high frequencies), and is an excellent general reference on much of the work in Part 1 of this book. Horowitz and Hill (1989) provides a detailed practical treatment of much of the work of Part 1, and in addition describes a range of electronic devices and their characteristics: this book cannot be recommended too highly, but its strength lies in its practical detail, and it is rather limited on theory (i.e. it is more a practical reference than a textbook).

1.13 Problems

1. Derive equations (1.30–1.33) expressing the y-parameters in terms of the h-parameters.
2. Obtain equations expressing the BJT y-parameters directly in terms of the T-parameters, starting from first principles. Confirm that your equations give correct values of the y-parameters, using the values of the T-parameters given in Section 1.3.

3. Determine the output impedance of a common emitter amplifier (Figure 1.12) using an *h*-parameter model. Use your result to confirm equation (1.62).

4. Noting that the values of y_f for a BJT and an FET vary with output current I (I_c and I_d respectively), determine whether there is a current above which it is better to use the one device in preference to the other. Use the typical values of the BJT and FET parameters given in the text to give substance to your answer.

5. A variable phase-shifting circuit can be built by adding a variable resistor R in series with a fixed capacitor C to the phase-splitting circuit described in Section 1.6, R being joined to the emitter output and C to the collector output, the final output being taken from the connection between R and C. Show that this arrangement gives an output of constant amplitude and with phase varying from $0°$ to nearly $180°$. How close will the phase approach $180°$ at $1\,\text{kHz}$ if $R = 10\,\text{k}\Omega$ and $C = 0.1\,\mu\text{F}$?

2

Circuit Building Blocks

2.1 Introduction

In this chapter we shall examine a variety of useful transistor circuits and circuit configurations. First we consider means of boosting the gain of a circuit by employing pairs of transistors coupled together. This type of arrangement is called a Darlington or "super-alpha" pair.

2.2 Darlington and complementary Darlington connections

The basic idea is to take two transistors and to connect them in a special way to produce a three-terminal composite device of much larger current gain. If npn and pnp transistors can be used, there are four possible circuit arrangements with this property (Figure 2.1). Of the circuits shown, (a) and (b) are called "Darlington pairs", and (c) and (d) are called "complementary Darlingtons" as they use complementary transistors. It is easy to see which terminal must be the emitter of the composite transistor in each case – it must be the one passing the most current, since $I_e = I_b + I_c$. Careful inspection of the circuits also reveals that the composite transistor is in all four cases of similar type (npn or pnp) to the first of its two transistors: thus (a) and (d) are npn, while (b) and (c) are pnp. A simple calculation shows that for a Darlington pair of type (a) or (b):

$$\beta = \beta_1 + \beta_2(\beta_1 + 1) = \beta_1\beta_2 + \beta_1 + \beta_2 \approx \beta_1\beta_2 \qquad (2.1)$$

31

(a)   (b)   (c)   (d)

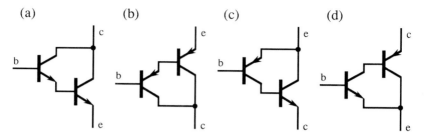

Fig. 2.1 Darlington and complementary Darlington circuits. Here (a) and (b) are Darlington pairs, and (c) and (d) are complementary Darlingtons. Note that in all cases the single equivalent transistor is of the same type (npn or pnp) as the *first* transistor of the pair.

while for a complementary Darlington:

$$\beta = \beta_1(\beta_2 + 1) = \beta_1\beta_2 + \beta_1 \approx \beta_1\beta_2 \tag{2.2}$$

Circuit (c) is highly useful on some integrated circuits, where it is straightforward to fabricate a low-gain pnp transistor but difficult to fabricate a high-gain one directly. However, in many applications such as hi-fi, a matched pair of high-gain npn and pnp transistors is required. For this purpose circuits (a) and (c) are often used together.

2.3   White emitter follower

Next we consider the White emitter follower (Figure 2.2). This circuit is useful when feeding a low-impedance load and a fast response is needed for positive *and* negative-going waveforms. The lower transistor operates between the output connection and the ground, and is capable of pulling the output voltage down rapidly. The upper transistor operates between the output connection and the upper voltage rail, and is capable of pulling the output voltage up rapidly. This second transistor acts as an emitter follower circuit, while the lower transistor acts as a normal common emitter amplifier. However, a complication arises since the lower transistor inverts the input signal: to overcome this problem, the lower transistor is fed from the collector of the upper transistor. Various capacitors and resistors are also required in order to ensure that both transistors operate under appropriate conditions. To avoid some of these complications, an alternative version of the circuit was devised, using a complementary pair of transistors (Figure 2.3). Notice

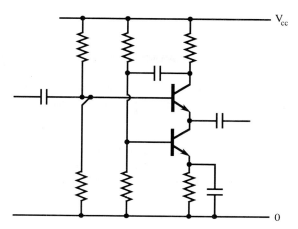

Fig. 2.2 White emitter follower.

that in reality this is an emitter follower constructed using a slightly modified complementary Darlington. Unfortunately, the significantly lower complexity of the circuit also reflects the fact that it only reacts rapidly when pulling the output voltage up, unlike the original White emitter follower. The complementary emitter follower aims to overcome this shortcoming and is described next.

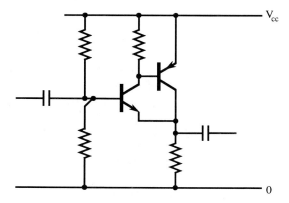

Fig. 2.3 Modified White emitter follower. This circuit is a White emitter follower modified so as to cut down the number of coupling capacitors and bias resistors. However, closer scrutiny shows that it can also be regarded as an emitter follower constructed using a slightly modified complementary Darlington.

2.4 Complementary emitter follower

In many applications such as audio amplification, "push-pull" power output stages are needed, and for this purpose the White emitter follower is inadequate since it is not symmetric: note, for example, that the upper and lower output connections have different intrinsic output impedances and speeds of operation. In particular what is needed is a symmetric circuit which produces no so-called "crossover distortion" – i.e. one with a perfectly symmetric linear voltage transfer function. The complementary emitter follower circuit (Figure 2.4) is symmetric and therefore there should be no even harmonic distortions in the output waveform. However, there is a range of input voltages $\sim 2V_{be}$ ($\approx 1.2\,\mathrm{V}$) over which both transistors are non-conducting. To understand this, note that when there is no a.c. input voltage, the base and emitter voltages are all, by symmetry, equal to $V_{cc}/2$, and both transistors are therefore cut off. The upper transistor only starts conducting when the a.c. input voltage pulls its base up by an amount V_{be} ($\approx 0.6\,\mathrm{V}$), and the inverse situation applies for the lower transistor. In fact, this picture does not take account of the fact that the switch-on of each transistor is a gradual function of input voltage. Thus the true combined voltage transfer function may be depicted as in Figure 2.5(a). To tackle this problem special bias circuitry can be included (Figure 2.6), and this has the effect of sliding the component voltage transfer curves in Figure 2.5 relative to each other, until their sum gives a good approximation to linearity (see Figure 2.5(b)). In the circuit of Figure 2.6, linearity has been improved further by including

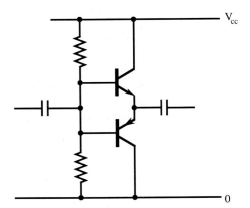

Fig. 2.4 Complementary emitter follower. This is an a.c. emitter follower circuit, using a complementary pair of transistors joined symmetrically to give a "push-pull" power output stage.

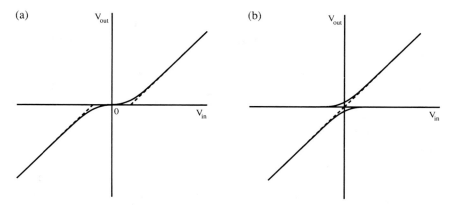

Fig. 2.5 Elimination of crossover distortion. (a) The voltage transfer characteristic for the circuit of Figure 2.4. The crossover distortion near the origin can largely be eliminated by sliding the characteristics for the individual transistors relative to one another, until their sum approximates closely to a straight line through the origin, as in (b). This is achieved using the modified bias circuit of Figure 2.6.

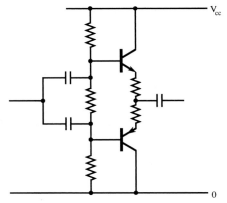

Fig. 2.6 Modified bias circuit for eliminating crossover distortion. Notice the small resistors in series with the emitters of the two transistors. These give extra linearity and thus help to eliminate crossover distortion.

a small "linearizing resistor" – typically in the range 1–$3\,\Omega$ – in series with each emitter.

Unfortunately, the general arrangement of Figure 2.6 is not satisfactory since a resistive circuit can only cancel the $2V_{be}$ offset accurately at a single temperature. Two diodes can be used in place of the resistive network, but it is more common to employ a "V_{be} multiplier" circuit, which can be used

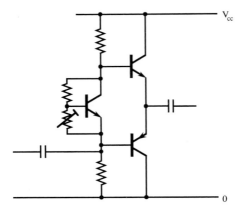

Fig. 2.7 More satisfactory bias circuit for the complementary emitter follower. In this case the bias circuit uses a V_{be} multiplier, as shown in more detail in Figure 2.8.

to cancel any multiple of V_{be}. The V_{be} multiplier will be described in detail below. Meanwhile, the complete circuit for a practical complementary emitter follower is shown in Figure 2.7. In Chapter 5 we shall see how a complete audio amplifier can be built with this type of output circuit.

2.5 V_{be} multiplier

This is a cunning circuit (Figure 2.8) with the property that the voltage V_o across it is a constant times the transistor V_{be} on-voltage ($\sim 0.6\,\text{V}$). The circuit adjusts itself, by a negative feedback process, so that $V_o \times R_2/(R_1 + R_2)$

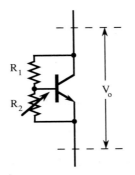

Fig. 2.8 V_{be} multiplier.

is just equal to V_{be}. Thus

$$V_o = \left(\frac{R_1 + R_2}{R_2}\right) V_{be} \qquad (2.3)$$

The reason for the name "V_{be} multiplier" is now obvious. The circuit has two notable features:

1. V_o may readily be adjusted to any convenient value by means of a single presettable potentiometer.
2. The temperature dependence of V_o is identical to that of the V_{be} of any transistor. For these reasons the circuit is ideal for cancelling out the V_{be} voltages of other transistors, as in the circuit of Figure 2.7.

An important factor in the use of the above circuit is its dynamic resistance ρ. This may be calculated by assuming that there is an a.c. voltage v_o across the circuit, deducing the current through R_2 and i_b, and then calculating i_c and the total a.c. current i_o. Thus we find:

$$\rho = \frac{R_1 R_2 + h_{ie}(R_1 + R_2)}{h_{ie} + (\beta + 1)R_2} = \frac{R_2(V_o/V_{be} - 1) + h_{ie}(V_o/V_{be})}{h_{ie}/R_2 + (\beta + 1)} \qquad (2.4)$$

where we have eliminated R_1 using equation (2.3). For any given V_o, this expression is clearly minimized by reducing R_2 as far as possible. However, below a certain point, most of the current would be flowing through the resistor chain, and on the contrary, we require the circuit to act as a chain of diodes: thus most of the current must come from the collector. For this to be so, R_2 must obey the condition:

$$R_2 \gg h_{ie}/(\beta + 1) \qquad (2.5)$$

while at the same time we require:

$$h_{ie} \gg R_2 \qquad (2.6)$$

so that equation (2.3) is accurate and the factor in brackets is independent of temperature.

Once R_2 has been decided, and V_o has been specified, ρ is completely determined by equation (2.4). Indeed, if we take R_2 to have the value $ch_{ie}/(\beta + 1)$, where $1 \ll c \ll \beta$, we finally obtain:

$$\rho = [(V_o/V_{be} - 1)c/(\beta + 1) + (V_o/V_{be})]h_{ie}c/(c + 1)(\beta + 1)$$

$$\approx (V_o/V_{be})h_{ie}/(\beta + 1) \qquad (2.7)$$

Thus ρ has the value $h_{ie}/(\beta+1)$ for $V_o = V_{be}$, and increases approximately linearly with increase of V_o. In fact, the a.c. voltage drop remains about the same proportion of the d.c. voltage drop as V_o increases – a quite reasonable result. It is of interest that equation (2.7) predicts values for ρ of 20, 40, 60 and 80 Ω at $V_o = 3$, 6, 9 and 12 V respectively for a BC107 at $I_c = 25$ mA: these values of ρ contrast with dynamic resistances 120, 27, 25 and 35 Ω for a typical series of Zener diodes with the same nominal voltages. Evidently, the V_{be} multiplier is of value mainly for producing relatively small voltage drops – *especially* when these are not integral multiples of 0.6 V. However, to obtain a voltage drop of 1.2 V, for a circuit such as that of Figure 2.7, R_1 and R_2 would typically both be $\sim 70\,\Omega$, and ρ would be about 8 Ω.

Before leaving this section, note that the predictions of ρ are based on the simplified model expressed by equations (2.5) and (2.6). However, the important point is that the design can be optimized: furthermore, there is a tradeoff in the design between dynamic resistance ρ (which should be low) and temperature dependence (which should be close to that of a diode). Clearly, there is more subtlety in this simple circuit than there first appears to be!

2.6   Diode-connected transistor

We next consider the diode-connected transistor shown in Figure 2.9. In fact this is a special case of the V_{be} multiplier discussed above, with $R_1 = 0$ and $R_2 = \infty$. Clearly, it has an output voltage equal to V_{be}, and in addition its I-V characteristic must be a $(\beta+1)$-amplified version of the emitter-base diode characteristic. Applying the V_{be} multiplier formula for ρ (equation 2.4)

Fig. 2.9   Diode-connected transistor.

in the specific case, we obtain the exact result:

$$\rho = h_{ie}/(\beta + 1) \qquad (2.8)$$

which gives $4\,\Omega$ for a BC107 at $I_c = 25\,\text{mA}$.

2.7 Bootstrapped emitter follower

It is the main purpose of an emitter follower circuit to act as a high-impedance buffer between two other circuits. Unfortunately, many emitter followers are unable to realize their potential in this respect, since the d.c. bias resistors used to stabilize the operation of the emitter follower themselves conduct substantial current, thereby constituting a significant load across the previous circuit (see also Section 1.7).

This effect can be cut down substantially by a clever trick. The output voltage variation is fed back to cancel the a.c. change in the bias voltage, thereby nearly eliminating the input a.c. current passing through the bias resistors. The complete circuit is shown in Figure 2.10. If the output voltage is v_o, the voltage across the series bias resistor R_s is $v_1 - v_o$ and the current through it is:

$$i_s = (v_1 - v_o)/R_s = v_1(1 - a_V)/R_s \qquad (2.9)$$

Hence the effective input impedance introduced by the bias resistor is:

$$Z_{\text{in,s}} = v_1/i_s = R_s/(1 - a_V) \qquad (2.10)$$

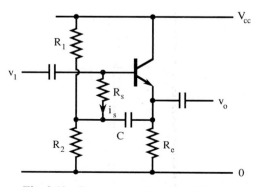

Fig. 2.10   Bootstrapped emitter follower.

Since a_V for an emitter follower is close to unity, $Z_{in,s}$ is many times larger than R_s. This means that a relatively small R_s, which will not reduce significantly the d.c. bias current provided by the main bias resistors R_1 and R_2, will be adequate to bring the emitter voltage input impedance back to quite a high value. For example, in a case where $R_e = 1\,k\Omega$, $R_1 = R_2 = 20\,k\Omega$, $R_s = 2\,k\Omega$, and $\beta = 200$, Z_{in} will be increased from $\sim 10\,k\Omega$ to $\sim 200\,k\Omega$ by this circuit arrangement.

2.8 Paralleled transistors

In certain applications, such as power amplifiers and power supplies, it may be necessary to exceed the current or power specifications of a given type of transistor. In that case it is attractive to attempt to put a number of such transistors in parallel. However, this can be a recipe for disaster, since if one transistor takes too much current and "blows", this will put more strain on the other transistors, and the remainder of them will be induced to blow (sometimes audibly, and at an accelerating rate!). However, this problem can be overcome by placing a low-value resistor in series with each of the emitters (Figure 2.11). Then an increase in current through any transistor will reduce the voltage across its emitter-base junction, and the current will be brought back towards the design level – with the result that the total current will be shared more evenly between the transistors. In fact, this is a further example of the use of local negative feedback – a topic that will be discussed in more detail in Chapter 5.

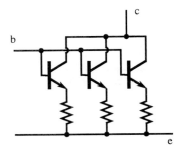

Fig. 2.11 Paralleled transistors. Here a small resistor (typically $\sim 1\Omega$) is placed in series with each emitter, thereby stabilizing the individual emitter currents. This is actually an example of negative feedback (see Chapter 5).

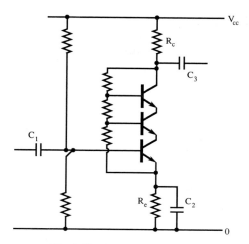

Fig. 2.12   Beanstalk amplifier.

2.9   Beanstalk amplifier

The aim of this circuit (Figure 2.12) is to amplify a.c. signals to amplitudes that are higher than any individual transistor can cope with. The idea is to spread the output voltage between a number of transistors, thereby preventing them from being exposed to unacceptable potential differences across their terminals. The lowest transistor and the resistor chain divide the collector voltage swing evenly between the set of transistors. For the circuit shown in Figure 2.12 containing three transistors, the maximum voltage swing hence approaches $3(V_{ce})_{max}$. To understand the mode of operation in more detail, imagine that the resistors are adjusted, in isolation, to give the ideal voltages at the collectors: clearly, they would need to have equal values to achieve this. Next, recall that each emitter-base junction will produce a voltage drop of 0.6 V. Thus the resistor chain will have to be adjusted to give a voltage at each base equal to the ideal collector voltage for the next lower transistor, plus 0.6 V in each case.

2.10   Artificial reactance circuits

In many applications, artificial reactances are required whose values can be adjusted easily over a considerable range – as for example in frequency meters or remotely controlled radio tuners. One of the main problems with variable capacitors is the small range over which they can be varied: this is because stray capacitances limit the range of capacitances to factors (C_{max}/C_{min}) typically between 10 and 20 (note that even smaller factors in frequency range result when these capacitors are used in tuned circuits, because of the square root relation $f = 1/2\pi\sqrt{LC}$). In addition, variable inductors are difficult to produce, and also large inductors and capacitors are bulky and expensive. Hence there is considerable merit in having circuits which can simulate inductors and capacitors, especially if these can be constructed conveniently using resistors and low-value capacitors, together with active circuit elements such as transistors.

The circuit of Figure 2.13 acts as an artificial inductor. To understand its operation, consider the various currents and voltages:

$$i_c = \beta i_b \tag{2.11}$$

$$v - v_{be} = R i_R \tag{2.12}$$

$$v_{be} = (i_R - i_b)/j\omega C \tag{2.13}$$

$$v_{be} = h_{ie} i_b \tag{2.14}$$

Eliminating i_R and i_b gives:

$$v = v_{be} + R(i_b + j\omega C v_{be})$$

$$= i_b(h_{ie} + R + j\omega C R h_{ie})$$

$$= i_c(h_{ie} + R + j\omega C R h_{ie})/\beta \tag{2.15}$$

$$\therefore \quad v/i_c \approx (R + j\omega C R h_{ie})/\beta \tag{2.16}$$

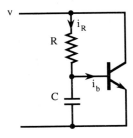

Fig. 2.13 Artificial reactance circuit. The circuit shown in this figure acts as an artificial inductor. Note that this is an idealized circuit and that d.c. bias circuitry is not included in the diagram.

where we have assumed that $R \gg h_{ie}$, which is normally valid for R greater than a few kilohms. This equation implies that the transistor is acting as an inductor of value CRh_{ie}/β in series with a resistance of value R/β, as suggested above. What is happening is that a small capacitive control voltage at the base, coupled with the natural inverting characteristic of the transistor, is making it act as an inductor. In a similar way, by interchanging the capacitor and the resistor, the transistor can be made to act as a capacitor. A formal proof of this is left as an exercise for the reader.

2.11   A square-law function generator

In this circuit (Figure 2.14) two matched FETs are joined together in a symmetrical way and fed with signals of opposite phase. (Note that signals of equal amplitude and opposite phase may be obtained with the 1-transistor phase-splitting circuit mentioned in Section 1.6, with $R_c = R_e$.) The FET is known to have an approximately square-law transfer function:

$$I_d = I_{dss}[1 + V_{gs}/V_p]^2 \tag{2.17}$$

where V_p is the "pinch-off" voltage.

In the given circuit, if the input voltages are applied with different phases relative to a constant voltage V_a, we have:

$$I_{d1} = I_{dss}[1 + V_{gs1}/V_p]^2 = I_{dss}[1 + (V_a + V_1)/V_p]^2 \tag{2.18}$$

$$I_{d2} = I_{dss}[1 + V_{gs2}/V_p]^2 = I_{dss}[1 + (V_a - V_1)/V_p]^2 \tag{2.19}$$

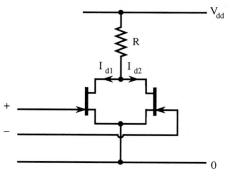

Fig. 2.14   FET square-law function generator. This square-law function generator circuit does not include d.c. bias circuitry. In addition, it has to be fed from a phase-splitter providing two waveforms of opposite phases at the two gate inputs.

Hence the output voltage in the given circuit is:

$$V_d = V_{dd} - R(I_{d1} + I_{d2})$$

$$= V_{dd} - 2RI_{dss}\left(1 + \frac{2V_a}{V_p} + \frac{V_a^2}{V_p^2} + \frac{V_1^2}{V_p^2}\right)$$

$$= V_k - kV_1^2 \tag{2.20}$$

where k and V_k are constants. In practice, voltage level setting circuits can eliminate V_k, and then the circuit becomes a true square-law function generator. Such circuits are useful for performing functions such as measuring power or forming part of analogue multiplying circuits, based on the "quarter-squares" principle, the latter being given by the following equation:

$$V_x V_y = \tfrac{1}{4}[(V_x + V_y)^2 - (V_x - V_y)^2] \tag{2.21}$$

Clearly, two square-law function generator circuits are required to build such a multiplier, and in addition an adder and two subtractors are required.

2.12 Summary

This chapter has described a number of useful circuit building blocks. These have included various emitter follower configurations (the White emitter follower, the complementary emitter follower, and the bootstrapped emitter follower), the Darlington and complementary Darlington configurations, and the V_{be} multiplier, most of which will prove useful later on when designing audio frequency and d.c. amplifiers and operational amplifiers (see Chapters 5 and 8). Other circuits discussed were paralleled transistors and the beanstalk amplifier – these being useful when circuits have to withstand currents or voltages greater than the rated values of the transistors to be employed – and artificial reactance circuits (useful, e.g., for generating frequency-modulated signals) and the square-law function generator.

Although some of these circuits might seem trivial at first, they embody important principles. In addition, a number of them illustrate the extra degree of freedom available to the designer when complementary devices (e.g. npn and pnp transistors) can be employed. This contrasts with the situation in the early days of radio when thermionic valves had to be used, and electrons were the only charge carriers that could be invoked.

Finally, although most of the circuits described in this chapter employ BJTs, in many cases equivalent versions of these circuits can be designed using FETs. Space precludes a full discussion of the situation, but note that the square-law function generator (Section 2.11) can be implemented *accurately* using FETs, whilst there is no obvious corresponding solution using BJTs.

2.13 Bibliography

As in the case of Chapter 1, the volumes by Calvert and McCausland (1978), Horowitz and Hill (1989) and Watson (1989), between them, provide significantly more depth on the work of this chapter. In particular, the second of these books provides a wealth of practical detail, while the third delves further into the underlying theory.

2.14 Problems

1. Give a full derivation of equations (2.4) and (2.7) for the dynamic impedance of a V_{be} multiplier. Also, obtain an alternative version of equation (2.7) in which R_2 is approximated as the geometric mean of h_{ie} and $h_{ie}/(\beta+1)$ in order to satisfy equations (2.5) and (2.6). Does this alternative equation predict significantly different values of ρ at voltages $V_o = 3, 6, 9$ and $12\,V$?
2. Use the theory of Section 1.7 to determine the basic voltage gain of an emitter follower. Hence obtain a corrected value for the input impedance of a bootstrapped emitter follower (see equation 2.10). How accurate are the numerical predictions of Section 2.7 for the increase in input impedance of an emitter follower produced by bootstrapping?
3. Following the methods of Section 2.10, find a formula for the effective capacitance of an artificial capacitor circuit. What *gain* in capacitance is achievable with this circuit, assuming reasonable resistor values? Show that this gain in capacitance could be improved considerably with an FET-based circuit.
4. It was assumed in Section 2.11 that identical FETs have to be used to form a square-law function generator. Show that this is not so, and that mis-matched FETs can be used if the input voltages are suitably adjusted.

What modifications must be made to the circuit of Figure 1.10 to provide a suitable phase-splitter to feed the FET circuit?

5. A *4-quadrant multiplier* is one which takes inputs of either sign and produces a product of the correct sign. Similar definitions apply for 2-quadrant and 1-quadrant multipliers. If a quarter-squares multiplier is built using the FET square-law circuit (see Section 2.11), in how many quadrants does it operate correctly?

3

Current Sources and Current Mirrors

3.1 Introduction

This chapter is concerned with constant current sources and their applications. On dealing with practical constant current sources in Section 3.2, it soon becomes apparent that the current mirror is a useful device, and it is studied in some detail in Section 3.3. Finally, the merits of the FET 2-terminal current source are considered in Section 3.4.

3.2 A basic form of constant current source

Figure 3.1 shows a commonly used form of constant current source. Circuits of this type can be very useful, e.g. in designing the linear ramp generator in an oscilloscope, where the circuit would feed a capacitor whose other terminal is connected to the V_{cc} power supply line (the basic principle is that the voltage across a capacitor being charged from a constant current source increases linearly with time).

Unfortunately, although Zener diodes can be fabricated with very low temperature coefficients, circuits such as that described above drift because transistor characteristics change markedly with temperature. This problem can be overcome by cancelling temperature drift by inserting a diode in the Zener diode leg of the circuit, either as well as or instead of the Zener diode.

In Figure 3.2 a diode-connected transistor is used for this purpose. Figure 3.3 shows exactly the same circuit, re-drawn for ease of analysis. We start by ignoring I_x, which is equal to the sum of the relatively small base currents of the two transistors; to this approximation, the collector current is equal

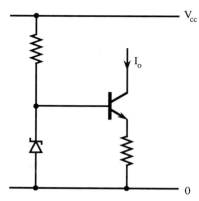

Fig. 3.1   Constant current source using a Zener diode.

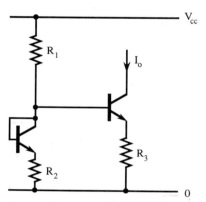

Fig. 3.2   Constant current source using a diode-connected transistor.

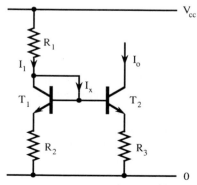

Fig. 3.3   Constant current source, redrawn in current mirror form.

to the emitter current for each transistor. Hence I_1 is given by:

$$V_{cc} - V_{be} = I_1(R_1 + R_2) \tag{3.1}$$

where V_{be} is *approximately* constant and is ~ 0.6 V.

We can now calculate I_o by assuming that the two emitter voltages are equal (both are about 0.6 V less than the common base voltage). Thus

$$I_1 R_2 = I_o R_3 \tag{3.2}$$

$$\therefore \quad I_o = \frac{(V_{cc} - V_{be})R_2}{(R_1 + R_2)R_3} \tag{3.3}$$

This equation is approximate because of our assumption about the collector and emitter currents of the two transistors. If $R_2 = R_3$, the equation for I_o simplifies to:

$$I_o = (V_{cc} - V_{be})/(R_1 + R_2) \tag{3.4}$$

Furthermore, if R_2 and R_3 are small,

$$I_1 \approx (V_{cc} - V_{be})/R_1 \tag{3.5}$$

where I_1 and hence I_o can be adjusted by adjusting R_1.

The circuit can now be considered as a current source feeding a "current mirror", with V_{cc} and R_1 forming the current source, and T_1, T_2, R_2, R_3 forming the mirror circuit. The latter *accepts* an input current I_1 and *generates* an output current I_o. We analyse the properties of current mirrors in the following section.

3.3 Current mirror circuits and their properties

Continuing with the train of thought at the end of the last section, we can postulate that a good current mirror is one for which:

1. I_o is closely equal to I_1.
2. Output impedence is high, i.e. the (parallel) output admittance is low.
3. There is good thermal stability (i.e. little current drift appears at the output).
4. A rather low voltage appears across the device (this is desirable so that as little a proportion of the power supply voltage as possible is wasted

in operating the circuit: note that the circuit of Figure 3.1 requires a minimum of about 4 V between output and ground – an obvious restriction on use of the circuit).

Aim 4 is rather restrictive, and implies that R_2 and R_3 should be eliminated. Since the previous analysis relied on the presence of these components, we now re-analyse the situation with reference to Figure 3.4. In an obvious notation,

$$I_1 = I_{c1} + I_{b1} + I_{b2} \tag{3.6}$$

$$I_o = I_{c2} \tag{3.7}$$

$$I_{c1} = \beta_1 I_{b1} \tag{3.8}$$

$$I_{c2} = \beta_2 I_{b2} \tag{3.9}$$

and if T_1 and T_2 are identical and at the same temperature:

$$I_{b1} = I_{b2} \tag{3.10}$$

and

$$\beta = \beta_1 = \beta_2 \tag{3.11}$$

$$\therefore \quad I_o = I_{c2} = \beta I_{b2} = \beta I_{b1} = I_{c1} \tag{3.12}$$

whereas

$$I_1 = I_{c1} + 2I_{c1}/\beta = I_{c1}(1 + 2/\beta) \tag{3.13}$$

$$\therefore \quad I_o = \frac{I_1}{1 + 2/\beta} \tag{3.14}$$

Since β is large, $I_o \approx I_1$, to a good approximation.

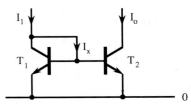

Fig. 3.4 Basic current mirror. This basic current mirror circuit has been simplified and idealized by eliminating the two emitter resistors. If implemented using discrete devices it will give poor performance, though it works well if fabricated on a single integrated circuit substrate.

The disadvantage of this circuit is that not only must the transistors be identical but also they have to be at identical temperatures, to make the approximations good ones. (Note in particular that emitter current varies *exponentially* with temperature – see equation (1.12) – thereby underlining the need for exact matching.) This is impossible to achieve unless the transistors are made simultaneously by the same fabrication process on the same substrate, almost in contact, so that they are subject to identical temperature variations*. Fortunately, integrated circuit technology permits this, and indeed current mirror circuits like that in Figure 3.4 are available commercially; these are encapsulated like transistors, and also have three connections.

Current mirror circuits can be made more accurate (I_o closer to I_1) by various techniques, one of these being to add an emitter follower, as in Figure 3.5(a). It will be clear that one reason why I_o cannot be exactly equal to I_1 is that the base currents for T_1 and T_2 are both derived from I_1; by inserting the emitter follower, $\beta + 1$ times less current is derived from I_1, so greater accuracy results. Similarly, the Wilson current mirror (Figure 3.5(b)) incorporates an emitter follower in a different way, but obtaining almost exactly the same improvement. In these more complicated circuits, the calculation of the ratios of the two currents is aided by noting various current values on the figure, starting with values of unity for the currents in the bases of the lowest two transistors, and then progressively multiplying by β, $1/(\beta+1)$ or $\alpha = \beta/(\beta+1)$, as necessary. The result for the emitter follower circuit of Figure 3.5(a) is:

$$\frac{I_o}{I_1} = \frac{\beta}{\beta + 2/(\beta+1)} = \frac{\beta^2 + \beta}{\beta^2 + \beta + 2} \tag{3.15}$$

while the ratio for the Wilson current mirror is:

$$\frac{I_o}{I_1} = \frac{(\beta+2) \times \beta/(\beta+1)}{\beta + (\beta+2)/(\beta+1)} = \frac{\beta^2 + 2\beta}{\beta^2 + 2\beta + 2} \tag{3.16}$$

Thus the latter gives only a marginal further improvement for moderate values of β. However, it has the added advantage of increasing the output impedance of the basic circuit by a factor $\sim (\beta+1)$. (This is an example of parallel current feedback – see Chapter 5.)

---

* For a full discussion of thermal drift in BJTs, see Chapter 8.

(a)

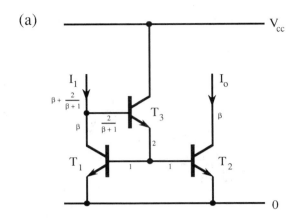

(b)

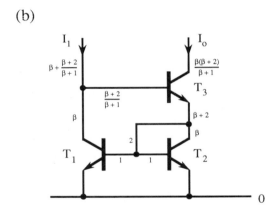

Fig. 3.5 Improved current mirror circuits. (a) A version of the circuit of Figure 3.4 which has been improved by the addition of an emitter follower. (b) The Wilson current mirror, which gives a marginal improvement over (a). Note that in more complex circuits such as these, insertion of relative current values, starting with $1+1$ at the bottom, facilitates calculation of I_o/I_1 (see text).

An interesting modification can be made to the current mirror circuit, by adding more transistors. Figure 3.6(a) shows a case where the output current is approximately $I_1/2$, and Figure 3.6(b) shows a case where the output current is approximately $3I_1$. Clearly, various integer fractions of the input current can be obtained in this way, but it must be emphasized that all the transistors have to be closely matched and should ideally be on the same substrate.

(a)

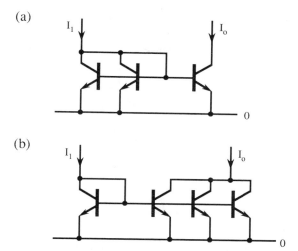

(b)

Fig. 3.6 Current mirrors giving different current ratios. (a) A circuit for which $I_o \approx I_1/2$, while (b) shows a circuit for which $I_o \approx 3I_1$. Note the simplified way of drawing transistors with common emitter and base connections.

There are many important applications of the current mirror. As indicated above, one is to produce a constant current source with a low offset voltage. Another is to "reflect" currents so that they appear (for example) between the output and ground instead of between output and the V_{cc} line. Figure 3.7 shows this situation arising in a voltage-to-current converter – a particularly important type of circuit.

A further application, used in most modern operational amplifiers, is to replace the resistor loads of a long-tail pair (T_1, T_2) by a current mirror (T_3,

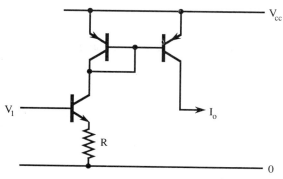

Fig. 3.7 Voltage-to-current converter.

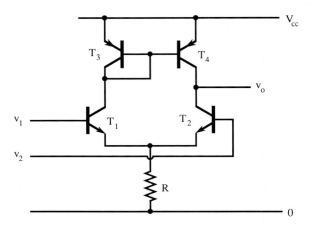

Fig. 3.8 Long-tail pair with active (current mirror) load. This circuit uses an active load to obtain high gain; in addition, use of a current mirror further doubles the gain, while at the same time achieving differential to single-ended conversion (giving an output voltage $v_o \propto (v_1 - v_2)$).

T_4), as in Figure 3.8*. If a single output is taken from the long-tail pair, the current gain produced by T_1 is not wasted but is added to that of T_2, thereby doubling the gain. This function of the mirror is called "differential to single-ended conversion". T_4 also acts as an "active" high-impedance load on T_2, thereby increasing the gain further. (A current source has a high-impedance output, but note that T_3 acts as a current *sink*† with a low impedance: this would typically have a value of just $4\,\Omega$ – see Section 2.6.)

3.4   An FET constant current source

It was noted in Section 1.7 that the FET differs from the BJT in a number of ways, one being its simpler self-biasing circuit. This turns out to be quite important when the FET is to be used as a constant current source, since

---

*For a full understanding of this circuit, the reader should refer to Chapter 8 which covers the long-tail pair.

† Some authors use the term *sink* to mean a current source of negative polarity. Here we retain the term *current source* for a high-impedance current source of *either* polarity: a *current sink* is then a device (essentially a short circuit connection) that is capable of accepting and dumping a current of either polarity.

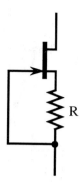

Fig. 3.9 FET constant current source. This FET constant current source is especially useful as it is self-biasing, unlike the corresponding BJT circuit. As a result, it becomes essentially a two-terminal device.

an adequate circuit can be built using just two components, as in Figure 3.9. Indeed, it is even possible to build a one-component constant current source by omitting the resistor! Such circuits are made into convenient two-terminal integrated devices and marketed as constant current sources – much as Zener diodes are sold as constant voltage sources. We can analyse the operation of the two-component circuit by first noting the equation for drain current:

$$I_d \approx I_{dss}\left(1 + \frac{V_{gs}}{V_p}\right)^2 \qquad (3.17)$$

Next by Ohm's law:

$$I_d R = -V_{gs} \qquad (3.18)$$

Eliminating V_{gs} now gives:

$$I_d \approx I_{dss}\left(1 - \frac{I_d R}{V_p}\right)^2 = I_{dss}\left(1 - \frac{2I_d R}{V_p} + \frac{I_d^2 R^2}{V_p^2}\right) \qquad (3.19)$$

Approximating this quadratic equation to the case of small R (corresponding roughly to a one-component circuit with remanent ohmic source resistance) and taking the appropriate solution, we find:

$$I_d \approx I_{dss}(1 - 2RI_{dss}/V_p) \qquad (3.20)$$

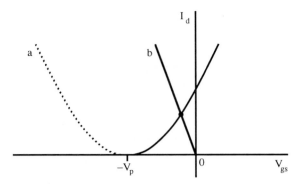

Fig. 3.10 Graphical solution for the FET constant current source. This diagram shows how the current provided by the FET constant current source of Figure 3.9 may be estimated graphically. Curve (a) shows the FET I_d v. V_{gs} square-law characteristic (equation 3.17), and line (b) is the load-line (equation 3.18) for the source resistor R. While two solutions are mathematically possible, the correct solution is the one that is relatively close to the origin.

Graphical solutions to these equations are perhaps easier to understand (see Figure 3.10).

3.5 Summary

This chapter has explored transistor circuits that can act as constant current sources. The current mirror emerged naturally from the analysis, but it was demonstrated to be almost valueless without the two emitter resistors – *except* when fabricated on one i.c. substrate. Various current mirror circuit configurations were shown to give improved accuracy. In addition, the current mirror was seen to lead to the possibilities of (a) active loads and differential to single-ended conversion for long-tail pair amplifiers, and (b) voltage-to-current converters. It is interesting to notice that when a current mirror is used as an active load, the input transistor exhibits a very low impedance, whilst the output transistor exhibits a high impedance: it is the latter which leads to the possibility of high gain when used as a collector load, whether in a long-tail pair or in a common emitter amplifier. Further illustrations of these concepts will appear in Chapter 8.

3.6 Bibliography

As in previous chapters, the reader will benefit by referring to the volumes by Calvert and McCausland (1978), Horowitz and Hill (1989) and Watson (1989). In addition, current mirrors have been covered quite well in for example Ritchie (1987) and a number of articles in *Wireless World* (e.g. Hart, 1970; Lidgley, 1979; Wilson, 1981, 1986). In fact, *Wireless World* has maintained a high level of informative articles on various novel aspects of electronics over quite a long period of time (certainly since the 1950s), and is worthy of serious attention. For more information on the widely used Wilson current mirror, see the original article (Wilson, 1968).

3.7 Problems

1. Find the ratio of I_o/I_1 for the current mirror circuit shown in Figure 3.11. Compare the effectiveness of this circuit with that of the emitter follower current mirror and the Wilson current mirror (Figure 3.5).

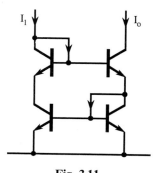

Fig. 3.11

2. Derive an exact formula for the current provided by an FET constant current source (Figure 3.9). If I_d is to be equal to $I_{dss}/2$, what value must R be given?
3. Draw the circuit of a common emitter amplifier using an active load incorporating a current mirror. Determine the gain achievable by such a circuit, in terms of h-parameters. By what factor is the gain improved relative to the gain of a basic common emitter amplifier?

4

Common Base and Cascode Amplifiers

4.1 Introduction

This chapter describes two types of amplifier that are widely used for radio-frequency work. Both types of amplifier are particularly valuable at these high frequencies because they overcome an important problem that arises with the usual common emitter type of amplifier: this problem is known as the Miller effect. First, we consider the common base amplifier; then we study the Miller effect; finally, we see how the common base amplifier and a more complex circuit known as the cascode help to eliminate the Miller effect.

4.2 The common base amplifier

In the common base (CB) amplifier (Figure 4.1), the input voltage appears at the emitter, and the base is held constant, i.e. at ground potential. As V_e increases, V_{be} decreases, I_b is reduced, I_c is reduced, and V_c rises: thus the voltage gain is *positive*, unlike the situation for the common emitter (CE) amplifier (see Figure 1.12). Next, we show that the CB amplifier nevertheless has the same voltage gain, numerically, as the CE amplifier.

In a CE amplifier, a certain v_{be} generates a certain i_b, which in turn produces a certain i_c. Feeding this current into R_c results in an output voltage v_o. In the CE amplifier, $v_1 = v_{be}$, and we have:

$$v_o = -R_c i_c = -R_c h_{fe} i_b = -R_c \left(\frac{h_{fe}}{h_{ie}}\right) v_1 \tag{4.1}$$

$$\therefore \quad a_V = -R_c\left(\frac{h_{fe}}{h_{ie}}\right) = -R_c y_{fe} \qquad (4.2)$$

In a CB amplifier, $v_1 = -v_{be}$, and this is the only change in the situation. Thus

$$v_o = -R_c i_c = -R_c h_{fe} i_b = -R_c\left(\frac{h_{fe}}{h_{ie}}\right)(-v_1) \qquad (4.3)$$

$$\therefore \quad a_V = R_c\left(\frac{h_{fe}}{h_{ie}}\right) = R_c y_{fe} \qquad (4.4)$$

The fact that the two cases are numerically equal is ultimately because I_c (and hence v_o) is (to a first approximation) dependent only on v_{be} – it does not matter how v_{be} gets its value, whether by the base or emitter going up and down in voltage. Once v_{be} is determined, the value of i_c is immediately decided, via the equation:

$$i_c \approx y_{fe} v_{be} \qquad (4.5)$$

Although the CE and CB amplifiers have numerically equal voltage gains, their respective current gains are $i_c/i_b = \beta$ $(\gg 1)$ and $i_c/i_e = \alpha = \beta/(\beta+1)$ (≈ 1). Thus it is no surprise that their input and output impedances are quite different. For the CB amplifier, Z_{in} is very low and Z_{out} is very high – facts that can easily give the impression that the CE and CB amplifiers have totally different voltage gains. In addition, the intrinsic voltage gains are proportional

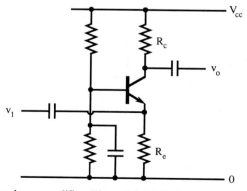

Fig. 4.1 Common base amplifier. Though initially it appears unusual, this common base amplifier has considerable similarities to the common emitter amplifier of Figure 1.12. This is because the d.c. biasing arrangements have to be nearly identical.

to $y_{fe} \approx 1/r_e \approx 40I_c$, which changes little with frequency; on the other hand, the CE current gain β varies rapidly with frequency, while $\alpha \approx 1$ and is necessarily almost independent of frequency. Thus at high frequencies, a CB stage can still give a high voltage gain, in spite of having a low current gain. Now the variations in gain of the BJT at high frequencies are normally modelled as due to the various internal capacitances*. As we shall see in the next section, the CB configuration is particularly attractive at high frequencies as it minimizes the effects of these capacitances.

4.3   The Miller effect

The Miller effect is the magnification of input capacitance as a result of the voltage gain of a circuit. It is a general effect which, in principle, applies to all amplifiers operating at high frequencies (e.g. more than 5 MHz) where stray capacitances become important. This means that it is best to analyse the general circuit shown in Figure 4.2, before coming to more specific conclusions.

The equations describing the operation of this circuit and giving its input impedance are:

$$Z_{in} = v_1/i = v_1/(i_1 + i_2) \tag{4.6}$$

$$i_1 = v_1 \left/ \left(\frac{1}{j\omega C_1} \right) \right. = j\omega C_1 v_1 \tag{4.7}$$

$$i_2 = (v_1 - v_o) \left/ \left(\frac{1}{j\omega C_2} \right) \right. = j\omega C_2 (v_1 - v_o) \tag{4.8}$$

$$i_3 = v_o \left/ \left(\frac{1}{j\omega C_3} \right) \right. = j\omega C_3 v_o \tag{4.9}$$

$$v_o = a_V v_1 \tag{4.10}$$

It turns out that equation (4.9) is not needed for the calculation of Z_{in}. Eliminating i_1 and i_2 in equation (4.6), we find:

$$Z_{in} = v_1/[j\omega C_1 v_1 + j\omega C_2 (v_1 - v_o)] = 1/[j\omega C_1 + (1 - a_V)j\omega C_2] \tag{4.11}$$

---

*Note that such models are only approximate and ignore charge carrier transit times which become especially important at very high frequencies.

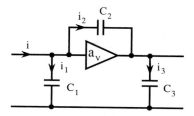

Fig. 4.2 General circuit for analysis of the Miller effect.

Thus, from the input the circuit appears to look like a capacitor* of value:

$$C_{\text{eff}} = C_1 + (1 - a_V)C_2 \qquad (4.12)$$

Evidently this equation will have serious consequences for an amplifier of large gain $|a_V| \gg 1$, because of the large capacitive load which then appears across the previous stage. Note that a CE stage suffers badly here, and a CB stage hardly at all; this is because the typical capacitances for a BJT have values:

$$C_{\text{be}} \approx 100\,\text{pF}$$

$$C_{\text{bc}} \approx 10\,\text{pF}$$

$$C_{\text{ce}} \approx 1\,\text{pF}$$

and for a CE stage $C_2 = C_{\text{bc}}$, whereas for a CB stage $C_2 = C_{\text{ce}}$. Of course, in both cases $C_1 = C_{\text{be}}$, and this is normally dominant for a CB stage:

$$(C_{\text{eff}})_{\text{CE}} = 100 + 10[1 - (a_V)_{\text{CE}}] \qquad (4.12a)$$

$$(C_{\text{eff}})_{\text{C}\beta} = 100 + 1[1 - (a_V)_{\text{CB}}] \qquad (4.12b)$$

Notice that the signs of a_V in the two cases are such that the terms are additive for a CE stage, while they tend to cancel for a CB stage – giving further reason for using the CB configuration to eliminate the Miller effect.

It turns out that the Miller effect can be even more serious than suggested by the above arguments. For suppose that the amplifier has a slightly inductive load: for example, in the case of Figure 4.3, the basic gain of the

---

* Clearly, our treatment has been simplified to ignore the normal input *resistance* of the amplifier, so the calculation gives a pure input capacitance.

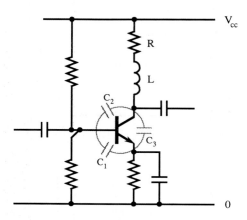

V_{cc}

Fig. 4.3 Common emitter amplifier with a slightly inductive load. When the Miller effect is present in an amplifier with a slightly inductive load, there is a risk that the input impedance of the amplifier will go negative, with the result that the circuit will oscillate at high frequency.

circuit becomes:

$$a_V' = -y_{fe}(R + j\omega L) \tag{4.13}$$

Then substituting a_V' for a_V in equation (4.11) gives the input impedance:

$$Z_{in}' = 1/\{j\omega C_1 + [1 + y_{fe}(R + j\omega L)]j\omega C_2\}$$
$$= 1/\{-y_{fe}\omega^2 LC_2 + [j\omega C_1 + (1 + y_{fe}R)j\omega C_2]\} \tag{4.14}$$

Clearly, the overall effect is to give the effective input capacitance already derived (as in equation 4.12), *and* a negative input resistance*. If this is large enough to counteract any remanent circuit resistance, then oscillations are likely to occur. Thus the Miller effect results in the whole circuit being potentially unstable.

There are a number of circuits which aim to overcome the Miller effect in one way or another. Perhaps the most obvious way of tackling the problem

---

*Zero resistance implies that a current can circulate for ever. Negative resistance implies that it can grow, and thus is a condition for oscillation at some frequency that will be determined by the various circuit reactances. (See also Chapters 5 and 6 for related positive feedback effects.)

is to "tune out" the feedback capacitance C_2, by inserting an inductor L_2 of suitable size in parallel with it (but in series with a blocking capacitor, in order to avoid upsetting the d.c. working conditions of the circuit). However, this solution is clumsy and a more elegant approach is the "neutrodyne" circuit. This employs a parallel-tuned circuit load with the d.c. power line joined to a tap on the inductor, while a neutralizing capacitor C_N feeds a signal of opposite phase to the input – thereby cancelling the effect of C_2 (Figure 4.4). In fact, this method is rather inflexible and unsatisfactory, since it works only over a restricted range of frequencies. Thus it is more common to employ a common base amplifier working on the principles explained earlier. However, the cascode connection, which incorporates a common base amplifier, is currently the most widely used and satisfactory means of tackling this problem (see below).

Finally, it is of interest that the neutrodyne of Figure 4.4 is essentially the same circuit as the well-known Hartley oscillator. If C_N is increased beyond the value required to neutralize C_2, it presents a short-circuit to high frequencies, and provides feedback with 180° phase-shift from collector to base. In conjunction with the 180° phase-shift of the transistor, this gives the overall positive feedback required to sustain oscillations at high frequencies (see Chapter 6).

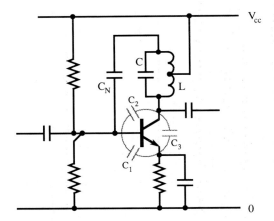

Fig. 4.4 The neutrodyne circuit. This circuit employs a tuned circuit collector load, with a centre tap joined to the V_{cc} line. Thus the alternate end of the inductor has an a.c. signal of opposite phase to that at the collector: feeding such a signal to the base via a "neutralizing" capacitor C_N can therefore cause the effect of the collector-base capacitance C_2 to be cancelled out. (For a centre-tapped inductor L, C_N must be equal to C_2.)

4.4   The cascode

Figure 4.5 shows a cascode amplifier. This is a 2-stage amplifier with a CE first stage and a CB second stage. In fact, the two stages are almost ideally harmonized to each other and thus achieve a good combination of properties for high-frequency amplification.

First, the CE stage achieves good voltage gain and good current gain (and therefore also excellent power gain), while its Miller effect is virtually eliminated since its collector is effectively prevented by the following CB stage from swinging up and down in voltage. The CB stage has this effect since, as stated earlier, it has a very low input impedance. (The reasons underlying this low input impedance will be explained in detail below.)

Meanwhile, the CB stage is able to achieve a large gain in voltage, which brings back the output voltage swing to the value it would ideally have had if the CB stage had not been present. (It is easy to see that this is so, since the CB stage transmits the CE collector current directly to the load resistor R_c.) Overall then, the CB stage has had the effect of almost eliminating the Miller effect for the CE stage. Sometimes, a final emitter follower stage is included with the cascode in order to give a low output impedance. Note that such a circuit will have used all three transistor configurations – CE,

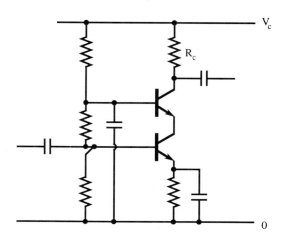

Fig. 4.5   The cascode. The cascode circuit shown here uses a common base (CB) amplifier as the load of a common emitter (CE) amplifier, with the result that the collector of the CE amplifier changes in voltage very little. This means that excessive charging and discharging of the collector-base capacitance is prevented and the Miller effect is virtually eliminated.

CB and CC (CC denotes *common collector* – which is an alternative description of an emitter follower).

It is worth considering the source of the low input impedance of a CB stage. In fact, to a good approximation the input section of a CB stage can be considered as the *output* of an emitter follower fed from a low source impedance. Now the emitter follower would have an output impedance equal to $h_{ie}/(\beta+1)$ in parallel with its emitter load (see Section 1.7). However, looking at the circuit from the emitter end, the input impedance (as it now is) appears merely as $h_{ie}/(\beta+1)$. Therefore, when the CB amplifier appears in the cascode circuit, the gain of the CE stage will be equal to:

$$a_V = -y_{fe} \times \frac{h_{ie}}{\beta+1} = -\frac{h_{fe}}{h_{ie}} \times \frac{h_{ie}}{\beta+1} = -\frac{\beta}{\beta+1} \approx -1 \qquad (4.15)$$

(This assumes that both transistors have the same value of h_{ie}.) Hence the effective input capacitance will actually be:

$$C_{eff} \approx C_1 + 2C_2 \qquad (4.16)$$

rather than the value given by setting $a_V = 0$ in equation (4.12). Clearly, it is wrong to suppose that the cascode eliminates the Miller effect – it merely reduces it to a low and manageable level.

Before leaving this topic, it is worth pointing out that the cascode is often implemented using FETs, or as a hybrid circuit having an FET high input impedance common source (CS) stage together with a BJT CB stage. In addition, dual-gate IGFETs present a good combination of properties for suppressing the Miller effect. The IGFET circuit (in Figure 4.6) is in fact a

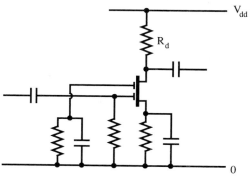

Fig. 4.6 Cascode circuit using a dual-gate IGFET device. Construction of a cascode amplifier is much simplified if a dual-gate IGFET is used, as in the circuit shown here.

form of cascode, in which the external connection between the two FET channels has been eliminated to make one composite device. It is interesting that this device is similar to the thermionic tetrode valve which was developed in the early days of radio, the additional grid between the first grid and anode acting as an electrostatic screen for reducing the capacitance between these two electrodes. In fact, the dual-gate IGFET circuit suggests that the cascode and the electrostatic screen approaches are fundamentally identical, just differing in detailed implementation!

4.5 Summary

There are necessarily three basic circuit configurations for three-terminal devices. For the BJT these are the common emitter (CE), common collector (CC) and common base (CB) configurations. The CE configuration is the best known, as it is the one giving the best overall combination of properties, including good voltage gain with reasonable input impedance. The CC configuration is the one practitioners generally learn about next – normally under the heading *emitter follower* – as having high input impedance and low output impedance. These two circuits are so much used that the practitioner soon accepts their properties intuitively. However, in this context the CB configuration tends to look somewhat unusual, and in any case it remains a rather specialized circuit that is used mainly at high frequencies.

Section 4.2 aims to show that the CB circuit has predictable properties, while Section 4.3 goes on to show the value of the CB amplifier in suppressing the Miller effect. Next, Section 4.4 shows how the CB amplifier is incorporated into the cascode circuit, which makes good use of its low input impedance characteristic, and at the same time capitalizes on its low Miller effect. Finally, the possibility of using an FET in the cascode – and in particular a dual-gate IGFET – is explored. Such possibilities form test cases for those who take the BJT to be pre-eminent amongst electronic amplifying devices. Suffice it to note here that much depends on the precise specification (including frequency range) of the circuit being designed, and the gains and bandwidths of any devices that might be used: what is a good solution at one frequency might well be mediocre at another frequency. In this context it is as well to note that the particular device and circuit to choose will also depend on expected signal levels and the required noise figure (see Chapter 11).

4.6 Bibliography

The topics in this chapter (especially various cascode circuits) are covered thoroughly, with relevant theory and detailed circuits, in Calvert and McCausland (1978). As ever, Horowitz and Hill (1989), is particularly useful and up-to-date on practical detail. In addition, Watson (1989) gives an excellent treatment of the variation of input impedance with load resistance, and the variation of output impedance with source resistance, for the CE, CB and CC configurations: he also shows clearly how these variations relate to each other in limiting cases.

4.7 Problems

1. Derive a formula giving the effective input impedance of a cascode constructed using (a) two JFETs; (b) an FET common source stage feeding a BJT common base stage. Follow the approach used in deriving equations (4.15) and (4.16).
2. Determine how the input impedance of a CB amplifier varies with the load resistor R_c. To what extent do your results modify the conclusions of Section 4.4?

5

Negative Feedback

5.1 Introduction

Figure 5.1 shows a rather general circuit which allows feedback to be applied from the output of an amplifier back to its input. Such circuits have a variety of useful and interesting properties, and it will be worth considering them carefully. In the following sections we analyse the properties of the general circuit, and then consider the consequences for practical amplifier circuits.

5.2 Analysis of a basic negative feedback configuration

In the circuit of Figure 5.1, the result of adding the feedback path is that the total voltage reaching the input of the amplifier becomes $v_1 + \beta v_o{}^*$. Hence:

$$(v_1 + \beta v_o)a_V = v_o \tag{5.1}$$

$$\therefore \quad v_1 a_V = v_o(1 - \beta a_V) \tag{5.2}$$

$\therefore$ the overall gain of the system is:

$$a_V' = \frac{v_o}{v_1} = \frac{a_V}{1 - \beta a_V} \tag{5.3}$$

---

*Note that it is usual to use β to denote the proportion of voltage fed back, even though this symbol is also commonly used for the BJT CE current gain.

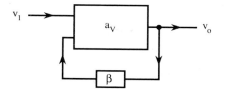

Fig. 5.1   General arrangement for negative feedback.

Defining the "loop gain" λ as the gain once right round the system, we have:

$$\lambda = \beta a_V \tag{5.4}$$

giving

$$a_V' = \frac{a_V}{1 - \lambda} \tag{5.5}$$

In Chapter 6 we shall explore what happens when β has such a sign that λ is positive, so that $|a_V'|$ can be greater than $|a_V|$. In this chapter we explore the case of negative feedback (n.f.b.), where $\lambda < 0$. In that case the total voltage reaching the input of the amplifier is reduced from v_1 to $(v_1 + \beta v_o)$, i.e. it is reduced by the factor

$$\zeta = \frac{v_1 + \beta v_o}{v_1} = \frac{v_o/a_V}{v_1} = \frac{a_V'}{a_V} = \frac{1}{1 - \lambda} \tag{5.6}$$

For n.f.b., $\lambda < 0$, but usually $|\lambda| \gg 1$ (i.e. $\lambda \ll -1$). Thus

$$0 < \frac{1}{1 - \lambda} \ll 1 \tag{5.7}$$

$$\therefore \quad |a_V'| \ll |a_V| \tag{5.8}$$

Also, $|v_1 + \beta v_o| \ll |v_1|$, which means that the two components at the input, v_1 and βv_o, almost exactly cancel. Evidently, this is what leads to the reduced voltage gain of the system. It also leads to a reduced current at the input terminal, since the voltage across the input impedance is no longer v_1, but $v_1 + \beta v_o$. Hence the input impedance is artificially increased by the factor $1 - \lambda$ which is normally much greater than unity:

$$Z_{in}' = (1 - \lambda) Z_{in} \tag{5.9}$$

Next, note that

$$a_V' = \frac{a_V}{1-\lambda} = -\frac{a_V}{\lambda} \bigg/ \left(1 - \frac{1}{\lambda}\right) = -\frac{1}{\beta}\left(1 + \frac{1}{\lambda} + \frac{1}{\lambda^2} + \ldots\right) \qquad (5.10)$$

Since $|\lambda| \gg 1$, we find that:

$$a_V' \approx -\frac{1}{\beta} \qquad (5.11)$$

The important point is that the overall gain is almost independent of a_V, so not only is the gain reduced but also it is stabilized. This implies it is independent of the specific active devices in the amplifier, and independent of temperature. Clearly it is important for the characteristics of the feedback connection to be designed carefully, as so much now depends on the stability of β.

5.3 Analysis of the effect of negative feedback on amplifier noise

To determine the effect of n.f.b. on noise generated by electronic components somewhere within an amplifier (or within the power supply feeding the amplifier), imagine it split into (or constructed from) two parts with a noise voltage v_N appearing between them, as in Figure 5.2. Then

$$[(v_1 + \beta v_o)a_1 + v_N]a_2 = v_o \qquad (5.12)$$

$$\therefore \quad v_1 a_1 a_2 + v_N a_2 = v_o(1 - \beta a_1 a_2) \qquad (5.13)$$

$$\therefore \quad v_o = v_1\left(\frac{a_1 a_2}{1 - \beta a_1 a_2}\right) + v_N\left(\frac{a_2}{1 - \beta a_1 a_2}\right) \qquad (5.14)$$

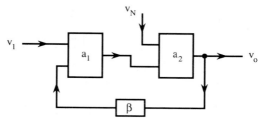

Fig. 5.2 Amplifier with noise input and feedback.

Recalling that the original amplifier had gain a_V, we have:

$$a_1 a_2 = a_V \qquad (5.15)$$

Thus it is clear that the v_1 term in the expression for v_o merely corresponds to the value for gain previously calculated. The v_N term tells us that the noise component at the output of the amplifier is just the non-feedback value $v_N a_2$, but reduced by the factor $1/(1-\lambda)$. Thus feedback has substantially improved the performance of the amplifier in this respect.

It is easy to show by similar arguments that mains hum or "drift" (in a d.c. amplifier) are correspondingly reduced, if generated (or otherwise injected) *within* an amplifier – but that no improvement can result if the disturbance is already present on the input signal. Distortion, if generated by non-linearities within an amplifier, is effectively noise, and again should be reduced by the factor $1/(1-\lambda)$. However, this argument has to be treated with caution, since the n.f.b. formulae we are quoting assume a *linear* amplifier. Thus this only gives an approximate answer, suitable for the reduction of small amounts of distortion but not (for example) gross overload.

5.4   Further consequences of negative feedback

Stability of gain following its reduction by n.f.b. has implications for frequency response, as shown in Figure 5.3. If a_V is dependent on frequency, then a_V' will tend to retain a constant value until a_V is so small that $a_V \approx a_V'$. This is so since $a_V' \approx -1/\beta$, though, as we have seen, this latter approximation is valid only when $|\lambda| = |\beta a_V| \gg 1$, i.e. when $|a_V| \gg |a_V'|$.

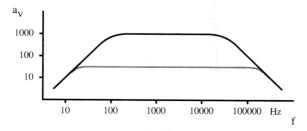

Fig. 5.3   Stabilization of gain by negative feedback. Here the black curve shows the original gain of an amplifier as a function of frequency, and the grey curve shows the gain after applying negative feedback. Notice that the effect of feedback is to stabilize the gain at a lower value over a significantly wider frequency range.

If the bandwidth of the original amplifier is limited by simple R-C circuits, then it is easy to show that the bandwidth is increased by a factor $(1-\lambda)$ on introducing n.f.b. However, it is significant that n.f.b. is unable to improve the gain–bandwidth product: this is a quantity that tends to remain unchanged, and is dependent on the type of electronic devices in use in the amplifier.

An important question in applying n.f.b. is whether to apply it separately to each stage of an amplifier, or to apply it to the overall system (Figure 5.4). In the first case the overall gain is:

$$a_1' = \left(\frac{a_V}{1-\beta_1 a_V}\right)^n \tag{5.16}$$

while in the second case it is:

$$a_2' = \left(\frac{a_V^n}{1-\beta_2 a_V^n}\right) \tag{5.17}$$

Differentiating these expressions with respect to a_V and dividing by the corresponding gain factors, and then simplifying, yields the two stability factors:

$$\frac{da_1'}{a_1'} = n\left(\frac{1}{1-\beta_1 a_V}\right)\frac{da_V}{a_V} \tag{5.18}$$

(a)

(b)

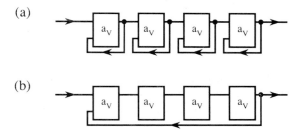

Fig. 5.4 Different ways of applying feedback in a composite amplifier. In (a) negative feedback is applied to each amplifier of a composite amplifier, while in (b) negative feedback is applied globally to the composite amplifier. In principle, the latter method is the more powerful (see text), though it tends to be more difficult to guarantee stability of such an arrangement because of the potentially greater number of phase changes within the feedback loop.

and

$$\frac{da_2'}{a_2'} = n\left(\frac{1}{1-\beta_2 a_V{}^n}\right)\frac{da_V}{a_V} \qquad (5.19)$$

$$\therefore \quad \frac{(da_2'/a_2')}{(da_1'/a_1')} = \frac{1-\beta_1 a_V}{1-\beta_2 a_V{}^n} \qquad (5.20)$$

Of course, the two composite amplifiers are to have the same overall gain, so $a_1' = a_2'$.

$$\therefore \quad 1-\beta_2 a_V{}^n = (1-\beta_1 a_V)^n \qquad (5.21)$$

i.e. β_1 and β_2 are adjusted to make these relations true. We now have:

$$\frac{(da_2'/a_2')}{(da_1'/a_1')} = \frac{1}{(1-\beta_1 a_V)^{n-1}} \qquad (5.22)$$

Since this factor is necessarily much less than unity, this means that *overall* amplifier feedback gives much greater stabilization of gain.

It has already been shown that n.f.b. reduces noise and distortion generated within an amplifier by a factor $1/(1-\lambda)$. We have also seen that it increases Z_{in}; in addition, it reduces Z_{out} by the same factor. Simple arguments may be used to confirm this.

Placing a load on the amplifier reduces v_o; this also reduces βv_o. However, βv_o initially almost cancelled v_1 and now this no longer happens; so v_o tends to increase again, to compensate. Eventually, βv_o will again cancel v_1 almost exactly, which means that v_o will go back to its original value – i.e. that which existed before the load was placed on the amplifier. This means that Z_{out} effectively has a rather low value.

We are gradually building up a picture of an n.f.b. amplifier as one which "resists change". Noise, drift, distortion, or the effects of loading are all reduced by the action of the feedback loop. In fact the purpose of the feedback path is to measure differences between the ideal output, $-v_1/\beta$, and the actual output, v_o, and to modify the input signal $(v_1 + \beta v_o)$ to make them as closely identical as possible. This strategy will naturally only be as accurate as $-v_1/\beta$ can be made to the ideal output voltage. For this reason it is crucial to make β exact, and this means it must be a ratio of resistor values: unless frequency dependence is required, the feedback path should ideally contain no capacitors or inductors.

5.5   Summary of the effects of negative feedback

The following list summarizes the effects of n.f.b. that have been covered so far:

1. It reduces and stabilizes a_V.
2. It increases Z_{in}.
3. It reduces Z_{out}.
4. It reduces noise, drift and distortion (increases linearity).
5. It increases bandwidth (but leaves gain–bandwidth product unchanged).
6. It has no effect on current gain a_i.

Note that these effects are to some extent related: e.g. 1 implies 5 (and to some extent 2, 3 and 4).

5.6   Some practical circuits incorporating negative feedback

This section describes a few circuits incorporating n.f.b. The first such circuit appears in Figure 1.10. There is no overt feedback loop in this case, but local n.f.b. is nevertheless being applied via the emitter resistor R_e. Thus the gain of the amplifier is reduced significantly, and the input impedance is increased significantly, as is seen from the analysis of Section 1.6, which (perhaps unexpectedly) applies for this circuit as well as for the emitter follower. However, the output impedance is not reduced as expected: this is because the feedback is not applied by direct sensing of the output voltage, but by more indirect means (actually this is an example of series current feedback: see below). Note that the input transistor is used to subtract the input signal and the feedback signal, by applying these on opposite sides of the emitter-base junction – a somewhat subtle design technique but one that is commonly used, as we shall see again below.

Figure 5.5 shows a 2-transistor circuit incorporating a definite feedback connection. In this case the connection is taken direct from the output, so we would expect that the output impedance is reduced in this case. Note that the feedback resistor R_f can conduct d.c. current between the two amplifier stages, and this means that the circuit is best designed in such a way as to equalize the d.c. voltages at either end of R_f (unless R_f is quite large). One way of ensuring this is to add a further transistor – as in the widely used "ring-of-three" amplifier shown in Figure 5.6.

Finally, Figure 5.7 shows, in outline, an audio frequency amplifier designed to feed a $16\,\Omega$ loudspeaker, and incorporating n.f.b. over three stages. This circuit embodies a number of circuit techniques that have been described in detail in Chapter 2. What at first appears to be positive feedback from the

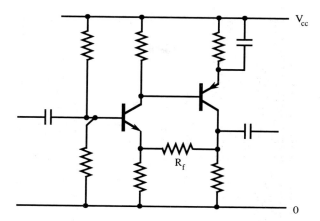

Fig. 5.5 Two-transistor feedback amplifier.

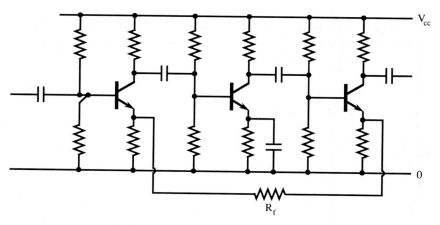

Fig. 5.6 Ring-of-three feedback amplifier.

output to the driver stage is better considered as bootstrapping* from the output to the driver load resistor in order to raise its effective a.c. impedance: this ensures greater linearity in driving the bases of the two output transistors (effectively, it engineers a current source drive for the V_{be} multiplier).

---

* Indeed, bootstrapping can consistently be defined as the stable application of positive feedback, keeping voltage gain and loop gain both slightly less than unity, with a view to increasing the apparent values of certain critical circuit impedances. However, the term originated as the idea of "pulling oneself up by one's own bootstraps!"

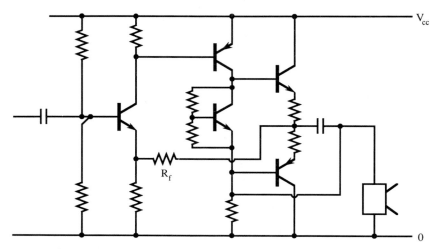

Fig. 5.7 Audio frequency amplifier incorporating negative feedback. This amplifier incorporates negative feedback over three stages, via R_f. It also incorporates bootstrapping between the output and the driver (second) stage in order to improve linearity when the latter feeds the bases of the two output transistors. Note the subtle combination of npn and pnp transistors.

5.7 Other types of negative feedback

The discussion of Sections 5.2–5.5 relates to a particular type of feedback called *series voltage* feedback: this means that the feedback signal is proportional to the output voltage v_o and is a voltage that is applied additively in series with the input voltage v_1. There are three other types of n.f.b.: *series current* (in which feedback is proportional to the output *current*), *parallel voltage* (in which the feedback is a current, proportional to the output voltage, that is applied in *parallel* with the input current) and *parallel current* (in which the feedback is a current, proportional to the output current, that is applied in parallel with the input current).

It can be shown that the two forms of voltage feedback have the effect of decreasing the output impedance, whereas the two forms of current feedback increase the output impedance. On the other hand, the two forms of series feedback increase the input impedance, while the two forms of parallel feedback have the opposite effect. Thus there is quite a range of circuit properties that can be obtained by careful choice of feedback paths.

Returning to the three circuits described in Section 5.6, we see that the first (Figure 1.10) and third (Figure 5.6) actually employ series current

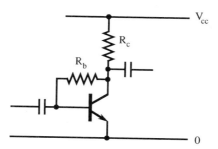

Fig. 5.8 Use of feedback to stabilize the operating point of a transistor. In this circuit, the collector is guaranteed to be at a d.c. voltage between 0 and V_{cc} for all possible values of β and R_b. However, this is achieved by a negative feedback configuration which also cuts down the a.c. gain.

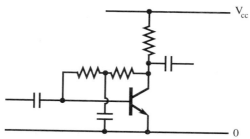

Fig. 5.9 Stabilization of operating point without reduction in gain. This version of the circuit of Figure 5.8 eliminates the a.c. component of the negative feedback, and hence preserves the intrinsic gain, while still being effective in stabilizing the d.c. operating point of the transistor.

feedback, though the second circuit (Figure 5.5) employs series voltage feedback. (However, note that the third circuit also becomes a series voltage feedback circuit if the output is taken from the final emitter.) The circuit shown in Figure 5.8 shows how n.f.b. is applied in order to stabilize the working point of a simple one-transistor amplifier, the bias resistor being joined from base to collector rather than direct to the positive power supply line. The feedback cuts down the a.c. gain of the amplifier, and reduces the input impedance, unless steps are taken (see Figure 5.9) to prevent these effects. The degree of d.c. stabilization afforded by this technique is not great* and it is not very widely used.

---

*However, it may be shown that the circuit is *guaranteed* to put any transistor into a working point somewhere between cut-off and saturation, whatever its value of β and whatever the size of the bias resistor R_b.

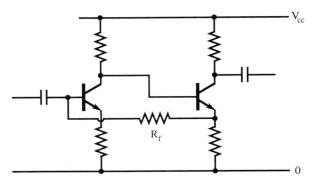

Fig. 5.10 Use of parallel current feedback to stabilize current gain.

Finally, a further interesting circuit is the amplifier shown in Figure 5.10. This also incorporates n.f.b., this time parallel current feedback. The effect of this is to reduce the input impedance and to increase the output impedance. It also reduces and stabilizes the *current* gain, so it is useful in linearizing the circuit as a current amplifier. In these respects it contrasts with properties 1–3 for series voltage feedback (see Section 5.5). The amplifier shown in Figure 5.10 might be of value for example in driving a chart recorder coil.

5.8 Stability of feedback circuits

Negative feedback circuits are quite sensitive to phase changes in a.c. signals as they progress though an amplifier. The reason for this is that a connection that is introduced to provide feedback may provide n.f.b. at one frequency and positive feedback at another frequency. In the latter eventuality, the circuit will almost certainly oscillate at some frequency for which there is an extra* phase change of 180° around the feedback loop. In some circumstances where the loop gain is insufficient to cause oscillation, it will nevertheless introduce instability and temporal distortion of any input signals.

These problems only arise (a) if the phase change that arises on going once around the complete feedback loop can approach some multiple of 360°, and (b) if the loop gain at this point is greater than unity. Clearly, a single *R-C* circuit can only produce up to 90° phase shift, and cannot cause problems. Unfortunately, in a circuit such as that in Figure 5.6, there are

---

*I.e. relative to the 180° phase-shift required to produce n.f.b.

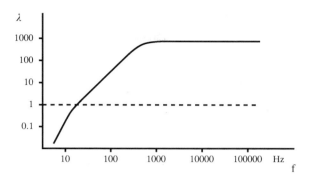

Fig. 5.11 Maintaining stability of an amplifier with feedback. Here one of the *R-C* circuits in an amplifier has been made dominant, so that the loop gain λ at the second corner frequency is less than unity: this ensures that the amplifier cannot oscillate at the frequency where the feedback becomes positive. Similar considerations apply at the high end of the frequency scale as well as the low end, as illustrated here.

several capacitors, each capable of producing $90°$ phase shift. However, oscillation and instability can be prevented by making one of the *R-C* circuits dominant, so that it produces large attenuation before the other *R-C* circuits produce appreciable phase shift. Then the loop gain will be less than unity at the point where the overall phase shift is equal to $360°$, and oscillation will not be possible. This situation is indicated in Figure 5.11.

5.9 Summary

This chapter has outlined the principles of negative feedback. Although there are four main ways in which negative feedback can be applied – by taking the voltage or current output and feeding it back in series or parallel at the input – the chapter concentrated on series voltage feedback as giving the most generally useful combination of properties. These properties included stabilizing the voltage gain and eliminating frequency, noise and temperature-induced variations on it, and also increasing input impedance and reducing output impedance. Finally, the stability of negative feedback loops was discussed.

The reader should note that negative feedback is especially important in that it cuts right across normal detailed electronic circuit design, and in particular helps the final circuit to have an idealized specification that is approximately independent of the characteristics of the specific (often highly non-linear) devices used in the circuit. Thus the gain of an amplifier may be controlled, somewhat paradoxically, by the ratio of two resistors in the circuit rather than by the transistors that provide the gain of the basic amplifier. Clearly, this should help to eliminate, at source, the many distortions that amplifiers can add to a signal.

5.10 Bibliography

Calvert and McCausland (1978) and Watson (1989) give more detailed coverage of the subject of negative feedback than has been possible in this chapter. Martin and Stephenson (1973) discuss in some depth the topics of stability and compensation. For useful practical circuits, see Horowitz and Hill (1989).

5.11 Problems

1. Determine which of the four main types of n.f.b. should be used in (a) a voltage-to-current converter, and (b) a current-to-voltage converter.
2. Derive a formula giving the working point of the transistor shown in Figure 5.9. Show that the working point is always somewhere between cut-off and saturation, whatever the values of β, R_b and R_c.

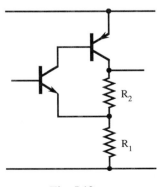

Fig. 5.12

3. Work out all the currents flowing in the modified emitter follower circuit of Figure 5.12, which uses a complementary Darlington. Deduce the voltage gain of the circuit. Show that for very large β current gains of the two BJTs, the voltage gain depends only on the ratio of the two resistors. Show that the final formula is consistent with the theory of Section 5.2 for a specific value of the voltage feedback factor β. Consider what happens as $R_2 \to 0$.

6

Sinusoidal Oscillators

6.1 Introduction

Oscillators operate by applying positive feedback so that a noise signal or small existing oscillation is repeatedly amplified until it becomes a large continuous oscillation. One of the important problems about using such a device as a source of sinusoidal waves is that once oscillations are established the loop gain has to be maintained at unity, or else the amplitude will grow until saturation effects cause gross distortion of the waveform. Hence a separate (negative) feedback loop has to be incorporated to stabilize the amplitude of oscillation. This is commonly achieved with the aid of a thermistor with a negative temperature coefficient of resistance.

6.2 Phase-shift oscillators

There are various types of sinusoidal oscillator. Perhaps the most obvious circuit employs an inverting amplifier and a $180°$ phase-shift network. A moment's thought will show that this requires more than two R-C circuits in cascade, or else a $180°$ phase-shift cannot be attained. Clearly, a 3-stage R-C network (Figure 6.1) will in principle suffice. However, there is a slight complication in that such a network is necessarily an attenuator, so the amplifier it is used with must have sufficient gain to bring the loop gain back to unity. We can estimate the required gain as follows. First assume that a single R-C stage is to give a phase-shift of $60°$, so that:

$$R \bigg/ \left(\frac{1}{\omega C} \right) = \tan 60° = \sqrt{3} \tag{6.1}$$

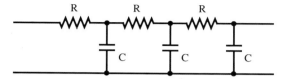

Fig. 6.1   Three-stage R-C phase-shift network.

Then its attenuation factor* will be:

$$L=\left[R^2+\left(\frac{1}{\omega C}\right)^2\right]^{1/2}\Bigg/\left(\frac{1}{\omega C}\right)=[1+(\omega CR)^2]^{1/2}=[1+(\sqrt{3})^2]^{1/2}=2 \quad (6.2)$$

This indicates that the attenuation factor for a 3-stage R-C circuit will be approximately 8. This result is approximate since each R-C stage loads the previous stage: in fact the true result is that the overall attenuation factor for 180° phase-shift is 29 (an exact calculation is left as an exercise for the reader). However, we can improve matters by using a 4-stage R-C network, in which case:

$$R\Bigg/\left(\frac{1}{\omega C}\right)=\tan 45°=1 \quad (6.3)$$

and the basic attenuation factor is then:

$$L=\sqrt{2} \quad (6.4)$$

This suggests that the overall attenuation factor will be approximately 4: in fact, an exact calculation shows that the overall attenuation factor is actually 18.4 – a very significant improvement on the result for the 3-stage network.

Although a 3-stage R-C network can be made to work satisfactorily, and a 4-stage R-C network should give a more practically useful oscillator, more often a slightly different strategy is adopted, and a *non*-inverting amplifier is used instead of an inverting amplifier. In this case a network with zero phase-shift is required. The well-known Wien bridge network (Figure 6.2) is suited to this type of application. We now analyse the Wien bridge circuit. For convenience we take the special case where the two resistors and the

---

* Attenuation factor L is defined as the inverse of gain G. Thus any attenuating network has $L \geqslant 1$, and $G \leqslant 1$.

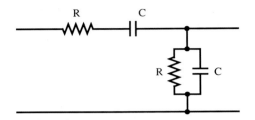

Fig. 6.2 Wien bridge R-C phase-shift network.

two capacitors are equal. Then

$$\frac{v_o}{v_i} = \frac{Z_o}{Z_i + Z_o} = \left(R \| \frac{1}{j\omega C}\right) \bigg/ \left[\left(R + \frac{1}{j\omega C}\right) + \left(R \| \frac{1}{j\omega C}\right)\right]$$

$$= \left(\frac{1}{R} + j\omega C\right)^{-1} \bigg/ \left[R + \frac{1}{j\omega C} + \left(\frac{1}{R} + j\omega C\right)^{-1}\right]$$

$$= \frac{j\omega CR}{[(1 + j\omega CR)^2 + j\omega CR]}$$

$$= \frac{j\omega CR}{[1 - (\omega CR)^2 + 3j\omega CR]}$$

$$= \left\{3 - \frac{j[1 - (\omega CR)^2]}{\omega CR}\right\}^{-1} \tag{6.5}$$

Clearly, this gives zero phase-shift for $\omega CR = 1$,

i.e.

$$\omega = 1/CR \tag{6.6}$$

At that frequency we have:

$$v_o/v_i = 1/3 \tag{6.7}$$

so there is an attenuation factor of 3 – considerably lower than for the 3- or 4-stage R-C networks – and most amplifiers will be able to provide the voltage gain of 3 that is necessary to sustain oscillations.

Another advantage of the Wien bridge oscillator is that varying the frequency of oscillation requires fewer components to be varied in unison. Typically, two resistors can be ganged together and adjusted, by factors in the range 10–50: greater variations in frequency can be achieved by switching in alternative pairs of capacitors – thereby permitting operation over many decades in frequency.

The Wien bridge oscillator is thus able to act as a very useful source of sinusoidal oscillations at frequencies up to $\sim 1\,\text{MHz}$ where stray capacitances and inductances start to become important. Instead of trying to eliminate such stray reactances, a better strategy for these higher frequencies is to make use of them, or at least to incorporate them into tuned circuits constructed using lumped inductors and capacitors. Then oscillators can be built which will operate satisfactorily up to $\sim 10^9\,\text{Hz}$, higher frequencies than this being obtainable using tuned microwave cavities (see Appendix C). However, space does not permit us to delve into a study of high-frequency tuned oscillators here.

6.3 Practical phase-shift oscillator circuits

Now let us return to the problem of building a non-inverting amplifier. A neat solution to this problem – and one which maps well to use of a Wien bridge – is the long-tail pair amplifier, used with single-ended output. Another possibility is the ring-of-three circuit described earlier. Figures 6.3 and 6.4

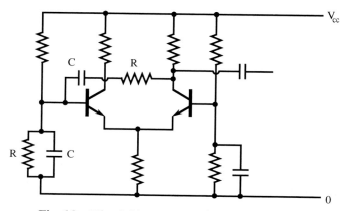

Fig. 6.3   Wien bridge oscillator using long-tail pair.

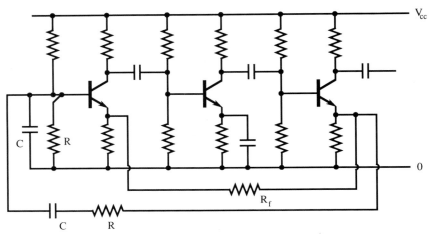

Fig. 6.4 Wien bridge oscillator using ring-of-three amplifier. In this circuit the resistors marked R and capacitors marked C are intended to control the frequency of oscillation: other capacitors should therefore have relatively large values so they do not substantially affect the phase of the waveform. Making R_f a thermistor with a negative temperature coefficient of resistance helps to stabilize the amplitude of oscillation.

show practical circuits constructed in this way. The latter circuit includes a negative feedback loop incorporating a thermistor with a negative temperature coefficient of resistance to stabilize amplitude of oscillation. Note that for amplitudes less than the operating amplitude, the loop gain of the oscillator is greater than unity, but it gradually drops to unity as the operating amplitude is approached. The degree of amplitude stabilization provided by the thermistor is more accidental than designed: hence an overt amplitude stabilization feedback loop is to be preferred. This may be achieved by measuring the signal amplitude in a separate circuit, and then feeding back the amplitude information to a linear device with variable voltage-controlled amplification. Fortunately, a junction FET can be used as a voltage-controlled resistor (VCR): for this purpose it is operated in its ohmic region – well away from the saturation region that is used in FET-based amplifiers (Figure 6.5). In the ohmic region the FET current–voltage characteristic is not exactly linear, but it can be linearized by means of a simple feedback circuit (Figure 6.6). Figure 6.6 also shows the extra components needed to control gain in an oscillator or other circuit, and Figure 6.7 shows how a complete oscillator is controlled in this way. Note the need for a time-constant to provide amplitude stabilization over many cycles of the a.c. waveform.

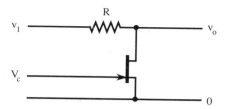

Fig. 6.5 FET as a voltage-controlled resistor. Here a FET is used as a voltage-controlled resistor (VCR): in conjunction with the resistor R, the FET is able to produce electronically controllable attenuation of the a.c. input voltage v_1.

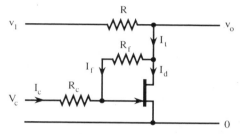

Fig. 6.6 Linearized version of FET voltage-controlled resistor. This linearized version of the circuit of Figure 6.5 employs negative feedback to cancel out the second-order V_{ds}^2 term from the FET conductance.

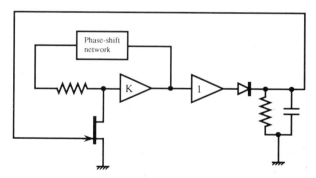

Fig. 6.7 Use of FET VCR to control the gain in an oscillator. This diagram shows in block form the circuit of an amplitude-stabilized feedback oscillator. A buffer amplifier is used to feed the rectifier diode so that the oscillator output is not unduly loaded and distorted. For clarity, the FET VCR is not the linearized form of the circuit. Note that d.c. level-shifting circuits will be required if the FET is to be properly (i.e. negatively) biased. For amplifier notation, see Figures 7.3 and 7.4.

To analyse the circuit of Figure 6.6 we can proceed to write down the five equations governing its operation, namely:

$$I_d \approx (2I_{dss}V_{ds}/V_p)[(1 + V_{gs}/V_p) - V_{ds}/2V_p] \tag{6.8}$$

$$V_c - V_{gs} = I_c R_c \tag{6.9}$$

$$V_{ds} - V_{gs} = I_f R_f \tag{6.10}$$

$$I_f = -I_c \tag{6.11}$$

$$I_t = I_d + I_f \tag{6.12}$$

We can then solve these to find how I_t depends on V_c and V_{ds}, and proceed to minimize the term in V_{ds}^2. However, it is useful to adopt a more intuitive approach. First, the resistor chain $R_c + R_f$ provides a constant contribution to the conductance, and so we have to examine the FET for any contributions which vary with V_{ds}. Second, the resistor chain $R_c + R_f$ acts as a potential divider, giving a modified voltage at the gate, equal to:

$$V_{gs} = (V_c R_f + V_{ds} R_c)/(R_c + R_f) \tag{6.13}$$

This has two effects, one being to introduce a further V_{ds}^2 contribution to I_d, via the existing V_{gs} term (see equation 6.8), and the second being to attenuate the control voltage by the factor $R_f/(R_c + R_f)$. We now attend to the first of these effects. If the two V_{ds}^2 contributions are to cancel out, we must have:

$$\frac{V_{ds}R_c/(R_c + R_f)}{V_p} = \frac{V_{ds}}{2V_p} \tag{6.14}$$

$$\therefore \quad 2R_c = (R_c + R_f), \quad \text{i.e.} \quad R_c = R_f \tag{6.15}$$

Thus, the non-linearity is completely eliminated by making these two resistors equal. However, this has not been achieved without cost, since the control voltage has been cut down by the factor $R_f/(R_c + R_f) = 1/2$. Fortunately, the FET has high input impedance, and so the value of $R_c + R_f$ can be raised to $\sim 1\,\mathrm{M}\Omega$: thus the parallel conductance of the resistor chain results in no significant further loss in effectiveness of the circuit.

Finally, it is worth remarking that the FET does not need to carry any d.c. current, and the drain circuit can be isolated from the signal line by a suitable blocking capacitor. This can give the resulting VCR circuit an odd appearance, but in fact it is useful in giving the designer additional freedom to fit the VCR into the remainder of the system.

6.4 Summary

This chapter has examined how oscillators are constructed by the action of positive feedback. For operation at moderate frequencies (up to $\sim 1\,\text{MHz}$), use of R-C phase-shift networks provides a useful strategy. With an inverting amplifier, a $180°$ phase-shift network is required, and with a non-inverting amplifier, a $0°$ phase-shifter is needed. 3- or 4-stage R-C networks giving $180°$ phase-shift attenuate signals excessively, and hence the simpler Wien-bridge $0°$ phase-shift network with an attenuation factor of 3 is much more convenient to use.

The chapter has also examined the amplitude of oscillation, and has shown that to avoid distortion a feedback amplitude-control loop is required. This may simply use a thermistor, or else (in more sophisticated designs) it may employ a voltage-controlled resistor (VCR) fed with a d.c. voltage from a rectified version of the original a.c. oscillation. This is a case where the FET comes into its own, since it can act as a highly effective VCR, and the latter operates even more effectively if linearized by a carefully adjusted local feedback loop. Note that the resulting circuit then has two negative feedback loops and one positive feedback loop, though the concepts involved are not at all difficult to understand.

6.5 Bibliography

For more detail on the topics of this chapter, the reader is again referred to Calvert and McCausland (1978), Horowitz and Hill (1989) and Watson (1989). Martin and Stephenson (1973) cover the theory of a number of types of oscillator, including both l.f. (e.g. phase-shift) and r.f. (e.g. Colpitts) types of circuit.

6.6 Problems

1. Derive a formula for the attenuation of a 3-stage R-C circuit, and confirm that when the overall phase change is $180°$, $\omega = 1/\sqrt{6CR}$, and the attenuation factor is 29.
2. Give a specification for the thermistor to be used in the oscillator circuit of Figure 6.4. (e.g. what value should it have, what should the tolerance on this value be, what should its temperature coefficient of resistance be, and to what extent do the values of these parameters depend on each other?)

7

Operational Amplifier Applications

7.1 Introduction

The operational amplifier is so called because it can be used to perform a variety of vital electronic operations merely by connecting it in various ways to basic components such as resistors, capacitors and diodes. What is almost universal in the use of operational amplifiers is the application of negative feedback. The reason for this is to approach some idealized performance criterion which makes use of the high basic gain of the operational amplifier.

The two main characteristics of the operational amplifier that must be borne in mind are its high voltage gain A and its high input impedance Z_{in}. Normally it is a reasonable approximation to take Z_{in} as infinite, while A will be at least 100,000. Most operational amplifiers possess two input connections and obey the equation:

$$v_o = A(v_+ - v_-) \qquad (7.1)$$

where the voltage gain A is taken as positive. First, we consider circuits for which the v_+ connection can be grounded and ignored.

7.2 Some basic circuits using operational amplifiers

A good proportion of operational amplifier circuits can be considered to have the form shown in Figure 7.1, where one or more components are

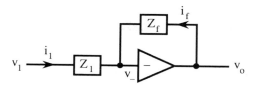

Fig. 7.1 Commonly used operational amplifier feedback circuit.

lumped into the two impedances Z_1 and Z_f. For such a circuit we have:

$$v_- = v_1 - i_1 Z_1 \tag{7.2}$$

$$v_o = -A v_- \tag{7.3}$$

$$v_- = v_o - i_f Z_f \tag{7.4}$$

$$i_f \approx -i_1 \tag{7.5}$$

this last equation only being valid since $Z_{in} \approx \infty$. Eliminating i_1 and v_- between equations (7.2), (7.4), (7.5), we get:

$$v_o = v_1 + i_f Z_f + i_f Z_1 \tag{7.6}$$

and applying equations (7.3), (7.4) to find i_f gives:

$$i_f = \frac{v_o(1 + 1/A)}{Z_f} \tag{7.7}$$

We may now eliminate i_f to finally determine the overall gain:

$$a_V = \frac{v_o}{v_1} = \frac{1}{1 - (1 + 1/A)(Z_1 + Z_f)/Z_f} = \frac{-Z_f}{(1 + 1/A)Z_1 + Z_f/A} \tag{7.8}$$

Clearly, if we may assume that $A \gg 1$, this equation simplifies considerably, giving:

$$a_V \approx -Z_f/Z_1 \tag{7.9}$$

The simplicity of this final expression, and its form, which is independent of the parameters of the operational amplifier, is one of the characteristics of circuits employing strong negative feedback. In addition, the fact that

$A \gg 1$ means (see equation 7.3) that:

$$v_- \approx 0 \qquad (7.10)$$

to a very good approximation. The negative input to the operational amplifier is thus called a "virtual earth", and many calculations are simplified by starting with equation (7.10) instead of equation (7.3). (Note that using this relation, equations (7.2) and (7.4) simplify considerably, and equation (7.9) becomes trivially deducible.)

The circuit of Figure 7.1 necessarily has an inverted input. In some applications it is useful to have a non-inverted input, and then the circuit shown in Figure 7.2 is employed. In this case equation (7.1) has to be used instead of equation (7.3), and in equation (7.2), $v_1 = 0$, giving:

$$v_- = -i_1 Z_1 \qquad (7.11)$$

Eliminating i_1, i_f and v_- between equations (7.1), (7.4), (7.5) and (7.11) now gives:

$$a_V = \frac{v_o}{v_+} = \frac{1 + Z_f/Z_1}{1 + (1 + Z_f/Z_1)/A} \qquad (7.12)$$

and if $A \gg 1$ we finally obtain:

$$a_V \approx 1 + Z_f/Z_1 \qquad (7.13)$$

and the virtual earth condition 7.10 is replaced by:

$$v_+ - v_- \approx 0 \qquad (7.14)$$

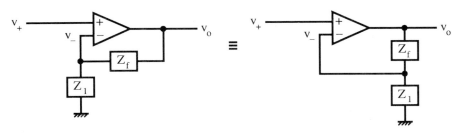

Fig. 7.2 Operational amplifier feedback circuit with positive gain. Note the two commonly used ways of drawing this circuit.

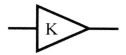

Fig. 7.3 Symbol for operational amplifier feedback circuit with gain K. For an amplifier of positive gain K, the circuit of Figure 7.2 can be employed, with appropriate resistor values given by equation (7.13).

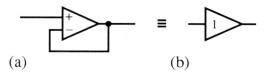

(a)   (b)

Fig. 7.4 Voltage follower circuit. This unity gain circuit, commonly used as a buffer amplifier, is constructed as in (a), but is often denoted as in (b).

The special case when Z_f and Z_1 are both resistors is especially useful, and for convenience this circuit is sometimes depicted as shown in Figure 7.3. In addition, the case when $Z_f = 0$, or $Z_1 = \infty$, or both of these conditions applies, leads to a_V being unity. In the last case the circuit (Figure 7.4(a)) is called a "voltage follower" and is a widely used form of buffer amplifier*. Sometimes the circuit is depicted as in Figure 7.4(b), in line with the notation of Figure 7.3.

7.3   First order low-pass filter

We now apply equation (7.9) to the case of a first-order low-pass filter (Figure 7.5). In that case we have:

$$a_V = -\frac{R\|\dfrac{1}{j\omega C}}{R_1} = -\frac{1}{R_1}\left(\frac{R\times\dfrac{1}{j\omega C}}{R+\dfrac{1}{j\omega C}}\right) = -\frac{R}{R_1}\times\frac{1}{1+j\omega CR} = -\frac{R}{R_1}\times\frac{1}{1+j\omega/\omega_2}$$

$$(7.15)$$

---

* A buffer amplifier is one which is capable of transparently driving one circuit from another: in general, this means that it has to have high input impedance, low output impedance and unit gain.

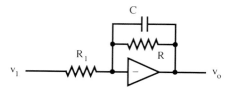

Fig. 7.5   First-order low-pass filter using an operational amplifier.

where $\omega_2 = 1/CR$ is the 3 dB* point ($|a_V|$ drops to $1/|1+\mathrm{j}| = 1/\sqrt{2}$, so $|a_V|^2$ drops to $\frac{1}{2}$, i.e. $-3\,\mathrm{dB}$).

When $\omega \gg \omega_2$,

$$a_V \approx -\frac{R}{R_1} \times \frac{1}{\mathrm{j}\omega/\omega_2} = \mathrm{j}/\omega CR_1 \qquad (7.16)$$

so $|a_V|$ drops off as $1/\omega$, and $|a_V|^2$ drops off as $1/\omega^2$, i.e. as $-6\,\mathrm{dB/octave}$.

In a second-order filter, in which two such amplifiers are connected in cascade, $|a_V|^2$ theoretically drops off at $\sim 12\,\mathrm{dB/octave}$, and in a third-order filter, $|a_V|^2$ drops off at $\sim 18\,\mathrm{dB/octave}$. However, this assumes that the individual amplifiers do not interact with each other – an assumption that is not in general a good one as the input and output impedances will vary with frequency (though it turns out to be adequate in certain practical cases). As a result, the optimal design of higher-order filters, and particularly those which are to have sharp cut-offs at particular frequencies, will not be an altogether trivial matter.

7.4   First-order high-pass filter

Figure 7.6 shows a first-order high-pass filter circuit, which we now analyse. In this case:

$$a_V = -\left(\frac{R_\mathrm{f}}{R + \dfrac{1}{\mathrm{j}\omega C}}\right) = -\frac{R_\mathrm{f}}{R} \times \frac{1}{1 + 1/\mathrm{j}\omega CR} = -\frac{R_\mathrm{f}}{R} \times \frac{1}{1 - \mathrm{j}\omega_1/\omega} \qquad (7.17)$$

where $\omega_1 = 1/CR$ is the 3 dB point.

---

* The decibel (dB) is a commonly used unit for expressing power ratios. The ratio of two powers P_1, P_2 is given in dB by the formula: $D = 10\,\log_{10}(P_1/P_2)$. Note that for power ratios of 2, 4, and 10, the respective values of D are 3, 6 and 10 dB, the first two of these values being approximate only (see Table 11.1). It is a useful feature of the decibel scale that raising a power ratio to power n *multiplies* D by n: this is important, for example, if first-order filters are cascaded to give higher-order filters, as suggested in the text.

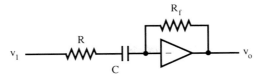

Fig. 7.6   First-order high-pass filter using an operational amplifier.

When $\omega \ll \omega_1$,

$$a_V \approx -\frac{R_f}{R} \times \frac{j\omega}{\omega_1} = -j\omega C R_f \tag{7.18}$$

so $|a_V|^2$ increases at $6\,\mathrm{dB/octave}$.

Again, it is possible to cascade stages in order to achieve higher rates of cut-off, but the same comments apply here as for low-pass filters.

7.5   Bandpass filter

We now consider a bandpass filter using an operational amplifier (Figure 7.7). In this case:

$$a_V = -\frac{Z_f}{Z_1} = -\left(\frac{R_2/j\omega C_2}{R_2 + 1/j\omega C_2}\right) \times \left(\frac{1}{R_1 + 1/j\omega C_1}\right) \tag{7.19}$$

$$= -\frac{R_2}{R_1}\left(\frac{1}{1 + j\omega C_2 R_2}\right) \times \left(\frac{1}{1 + 1/j\omega C_1 R_1}\right) \tag{7.20}$$

Substituting

$$x_1 = \frac{1}{\omega C_1 R_1} = \frac{\omega_1}{\omega} \tag{7.21}$$

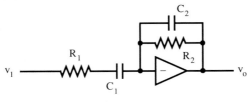

Fig. 7.7   First-order band pass filter using an operational amplifier.

and

$$x_2 = \omega C_2 R_2 = \frac{\omega}{\omega_2} \tag{7.22}$$

we get:

$$a_V = -\frac{R_2}{R_1}\left(\frac{1}{1+jx_2}\right) \times \left(\frac{1}{1-jx_1}\right) \tag{7.23}$$

$$= -\frac{R_2}{R_1}\left(\frac{1}{(1+x_1x_2)+j(x_2-x_1)}\right) = -\frac{R_2}{R_1}\left(\frac{1}{(1+\alpha)+j(\omega/\omega_2-\omega_1/\omega)}\right) \tag{7.24}$$

where

$$\alpha = x_1x_2 = \omega_1/\omega_2 \tag{7.25}$$

To find the bandwidth, we examine where $|a_V|$ drops to $1/\sqrt{2}$ of its maximum value. This occurs for:

$$1+x_1x_2 = \pm(x_2-x_1) \tag{7.26}$$

i.e.

$$1+\alpha = \pm(\omega/\omega_2-\omega_1/\omega) \tag{7.27}$$

Case 1. The first case of interest is when $\omega_2 \gg \omega_1$ $(\alpha \ll 1)$, in which case:

$$1+\alpha \approx \omega/\omega_2, \omega_1/\omega \tag{7.28}$$

i.e.

$$\omega \approx \omega_2(1+\alpha), \omega_1/(1+\alpha) \tag{7.29}$$

$$\therefore \quad \text{bandwidth} = \omega_2(1+\alpha)-\omega_1/(1+\alpha) = \omega_2(1+\omega_1/\omega_2)-\omega_1/(1+\omega_1/\omega_2) \tag{7.30}$$

$$\approx \omega_2+\omega_1-\omega_1(1-\omega_1/\omega_2) \approx \omega_2+\omega_1{}^2/\omega_2 \approx \omega_2 = 1/R_2C_2 \tag{7.31}$$

and midband gain

$$(a_V)_{\text{mid}} \approx -\frac{R_2}{R_1}\left(\frac{1}{1+\alpha}\right) \approx -\frac{R_2}{R_1} \tag{7.32}$$

Case 2. Another important case is when $\omega_2 = \omega_1 = \omega_0$, so that $\alpha = \omega_1/\omega_2 = 1$, $|a_V|$ drops to $1/\sqrt{2}$ of its maximum value for:

$$\omega/\omega_0 - \omega_0/\omega = \pm 2 \tag{7.33}$$

$$\therefore \quad \omega^2 \pm 2\omega\omega_0 - \omega_0^2 = 0 \tag{7.34}$$

giving

$$\omega = \omega_0(\sqrt{2} \pm 1) \tag{7.35}$$

where we have taken only positive values of ω.

$$\therefore \quad \text{bandwidth} = [(\sqrt{2}+1)-(\sqrt{2}-1)]\omega_0 = 2\omega_0 = 2/R_1C_1 = 2/R_2C_2 \tag{7.36}$$

Notice that the bandwidth in the latter case is about twice that in the former case. The reason for this is that in the latter case the response function does not reach the same height (see Figure 7.8), so technically the bandwidth is greater than it would otherwise be.

(a)

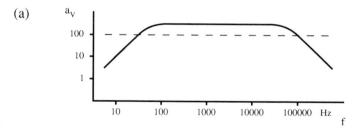

(b)

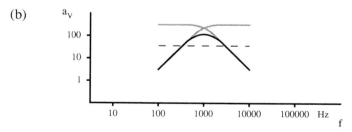

Fig. 7.8 Frequency response of band-pass filter. (a) The basic frequency response for a band pass circuit incorporating a low-pass and a high-pass element. (b) This shows how the maximum output level is reduced when the passband is small, the result, perhaps surprisingly, being that the bandwidth is double the initially expected value (see text).

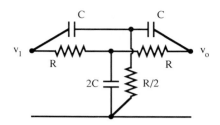

Fig. 7.9 Twin-tee band-reject ("notch") filter. This circuit can perhaps best be constructed in the laboratory by selecting *four* well-matched resistors and *four* well-matched capacitors.

It can easily be shown that the above circuit does not give a very sharply defined bandpass region (the attenuation drops off at $\sim 6\,\mathrm{dB/octave}$ in either direction outside the passband), and hence other circuits are commonly used to overcome this problem. The "twin-tee" filter is one that is widely used for the purpose. Figure 7.9 shows the basic circuit. It turns out that if the resistors and capacitors are well-matched as indicated, the filter tunes very sharply and theoretically has infinite impedance at the angular frequency:

$$\omega_0 = 1/CR \qquad\qquad (7.37)$$

This is perhaps an unexpected result, but one that is useful if a certain frequency has to be rejected. However, if we wish to build a bandpass filter, we simply have to insert this network into the feedback path of an operational amplifier (Figure 7.10). In practice this can cause problems, because there is nothing to limit the gain of the operational amplifier close to the resonance frequency – though this problem is easily overcome by placing a resistor R_f in parallel with the network as indicated by the dotted lines in Figure 7.10. Finally, note that a Wien bridge network can be used in a *positive* feedback

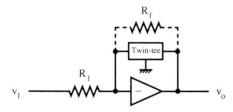

Fig. 7.10 Bandpass filter using a twin-tee circuit in a feedback loop. The inclusion of R_f (see dotted lines) permits the gain of the circuit to be limited to realistic values in the passband.

loop of an operational amplifier, with much the same effect; however, it is necessary to ensure that the loop gain does not exceed 3 or else the circuit will oscillate! (See also Section 6.2.)

7.6 The Sallen and Key filter

Sallen and Key investigated a whole class of second-order filters, one example of which is shown in Figure 7.11. We shall study this circuit in some detail as it illustrates some important principles. The equations governing the operation of the circuit are:

$$v_1 - v_2 = i_1 R \tag{7.38}$$

$$v_2 - v_3 = i_2 R \tag{7.39}$$

$$v_2 - v_o = (i_1 - i_2)/j\omega C_1 \tag{7.40}$$

$$v_3 = i_2/j\omega C_2 \tag{7.41}$$

$$v_o = K v_3 \tag{7.42}$$

Notice that by using the notation of Figure 7.3 we have saved working through a number of equations describing the details of how the positive gain amplifier is implemented. Understanding the operation of the circuit is now a question of eliminating v_2, v_3, i_1 and i_2 from the remaining five equations (7.38–7.42). After a fair amount of rather tedious manipulation we finally find:

$$a_V = v_o/v_1 = K/\{1 - \omega^2 C_1 C_2 R^2 + j\omega R[(1-K)C_1 + 2C_2]\} \tag{7.43}$$

Taking the case $K = 1$, this equation reduces to:

$$a_V = v_o/v_1 = [1 - \omega^2 C_1 C_2 R^2 + 2j\omega C_2 R]^{-1} \tag{7.44}$$

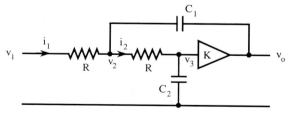

Fig. 7.11 Sallen and Key second-order low-pass filter.

The power response of this circuit is:

$$|a_V|^2 = [(1-\omega^2 C_1 C_2 R^2)^2 + (2\omega C_2 R)^2]^{-1}$$
$$= [1+(4C_2{}^2 R^2 - 2C_1 C_2 R^2)\omega^2 + (C_1{}^2 C_2{}^2 R^4)\omega^4]^{-1} \qquad (7.45)$$

Thus the circuit acts as a low-pass filter with unity gain at $\omega = 0$. If it is to have the flattest possible response in the pass-band, then we should make the ω^2 term in the square bracket vanish. This is achieved if:

$$C_1 = 2C_2 \qquad (7.46)$$

As an alternative, we could refrain from taking $K = 1$. In that case, the ω^2 term becomes:

$$\{[(1-K)C_1 + 2C_2]^2 R^2 - 2C_1 C_2 R^2\}\omega^2$$

Clearly, this can be adjusted to zero even when $C = C_1 = C_2$, provided that K obeys the relation:

$$(K-3)^2 - 2 = 0 \qquad (7.47)$$

i.e. $K = 3 \pm \sqrt{2} = 1.586$ or 4.414. The smaller of these values is normally taken, as then the (unaided) gain in the passband is closer to unity. However, what is important is that we now have two slightly different strategies for making the response flatter in the passband.

Thus we have shown that this type of Sallen and Key filter is able to act as a Butterworth filter where the latter is defined as having the response:

$$|a_V| = \frac{|a_{V0}|}{[1+(\omega/\omega_c)^2{}_n]^{1/2}} \qquad (7.48)$$

and n is defined as the *order* of the filter, and is normally equal to the number of reactances used in the circuit. The value of such Butterworth responses is that the first $2n-1$ derivatives of the function in the square brackets are zero, so that the response in the passband is *maximally flat* – i.e. is a maximum at $\omega = 0$, and remains nearly constant up to $\omega = \omega_c$. Outside this range the response curve plunges down at a very rapid rate, and for $\omega \gg \omega_c$ the response curve approximates to:

$$|a_V| = \frac{|a_{V0}|}{(\omega/\omega_c)^n} \qquad (7.49)$$

By contrast, the response curve for n uncoupled (i.e. buffered) R-C low-pass filter sections would be:

$$|a_V| = \frac{|a_{Vo}|}{[1 + (\omega/\omega_c)^2]^{n/2}} \qquad (7.50)$$

This curve has the same general monotonic shape, the same value at $\omega = 0$, and also identical behaviour (equation 7.49) for $\omega \gg \omega_c$: however, it is *far* inferior since the first $2n-1$ derivatives are not zero. Because of these important differences, the Butterworth paradigm is of some importance. It is therefore worth remarking how higher-order Butterworth filters can be constructed. In fact, this is highly straightforward, since higher-order filters can be built using a cascade of second-order filters, plus (in the case of odd n) a single first-order filter, with the total order summing to n. However, the composite second-order filters must have carefully adjusted gain values for the combined filter to have the ideal Butterworth response. The full theory of Butterworth filters and selection of gain values is beyond the scope of the present discussion, but appropriate gain values for various component first- and second-order filters are given in Table 7.1. (It is debatable whether it is ever worth using odd-order filters of this type, since the first-order sub-filter will require an operational amplifier, and once this is deployed it is trivial to enhance the circuit to the next even order.)

It is interesting that the Butterworth filter design philosophy described above is rather restricted. In particular, if a small amount of passband "ripple" is permitted, a significantly improved fall-off can be obtained outside the passband. We can see this with the simple Sallen and Key filter discussed earlier. If the second-order term is made to have the opposite sign to the fourth-order term, then the response in the passband will peak slightly at the border of the passband before plunging down more rapidly near $\omega = \omega_c$.

Table 7.1 Gain values for component first- and second-order filters.

n	First-order	Second-order		
1	1			
2			1.586	
3	1		2.000	
4		1.152		2.235
5	1	1.382		2.382
6		1.068	1.586	2.482

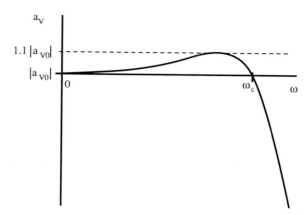

Fig. 7.12 Response for Chebyshev second-order low-pass filter. This figure shows the frequency response for a modified form of Sallen and Key filter: improved fall-off is obtained outside the passband by tolerance of 10% ripple within the passband.

Figure 7.12 shows this effect when the curve is adjusted to go to 10% above its original height near $\omega = \omega_c$. To achieve this we model equation (7.45) as:

$$|a_V|^2 = 1/W \tag{7.51}$$

where

$$W = 1 + a\omega^2 + b\omega^4 \tag{7.52}$$

$$\therefore \quad dW/d\omega = 2a\omega + 4b\omega^3 \tag{7.53}$$

which is zero for $\omega = 0$ and $\sqrt{-a/2b}$.

If the maximum of $|a_V|^2$ is to be 10% above the $\omega = 0$ value, we have:

$$1/1.1 = 1 + a(-a/2b) + b(-a/2b)^2 \tag{7.54}$$

$$\therefore \quad 0.1 = \frac{a^2}{4b} = \frac{(4C_2{}^2R^2 - 2C_1C_2R^2)^2}{4(C_1{}^2C_2{}^2R^4)} = \frac{(2C_2 - C_1)^2}{C_1{}^2}$$

so C_2 has to be increased or decreased by $\sim 5\%$.

This approach is adopted in the Chebyshev paradigm, and the amounts of ripple that are normally quoted are 0.5 dB and 2 dB (power variations of $\sim 12\%$ and $\sim 58\%$ respectively). Again, details of the theory underlying this approach are beyond the scope of the present work. However, it is worth noting that the Chebyshev paradigm is in general more realistic than the Butterworth paradigm, since the tolerances on the manufacture of electronic components are normally in the range 5–20%: thus Butterworth filters will in any case be subject to ripple because of component mismatch. Note that both of these types of filter suffer from poor phase characteristics – a property not analysed here. To overcome such problems, and give a fixed time delay through the circuit, the Bessel filter was devised. The reader should see Horowitz and Hill (1989), Kuo (1962) and the other references cited in Section 7.8 for details of such circuits which are especially important with pulses and digital signals.

Finally, although only low-pass filters have been discussed above, they may readily be transformed into high-pass filters: for example, in the circuit of Figure 7.11, if the two resistors R are interchanged with the two capacitors C_1, C_2, the resulting circuit becomes a high-pass filter, whose analysis is in close parallel with that above.

7.6.1 Equivalent *LCR* filters

It is of interest to compare the above types of filter with what can be achieved with the simple *LCR* type of filter. For generality we take the case where the inductor has a series resistance R_L (Figure 7.13). In that case the overall gain is:

$$a_V = \frac{v_o}{v_1} = \frac{1/j\omega C}{R_L + j\omega L + 1/j\omega C} = \frac{1}{1 - \omega^2 LC + j\omega C R_L} \tag{7.55}$$

$$\therefore \quad |a_V| = 1/[(1 - \omega^2 LC)^2 + \omega^2 C^2 R_L^2]^{1/2} \tag{7.56}$$

which is equal to $1/(1 + \omega^4 L^2 C^2)^{1/2}$ if:

$$2LC = C^2 R_L^2, \quad \text{i.e.} \quad L = \tfrac{1}{2} C R_L^2 \tag{7.57}$$

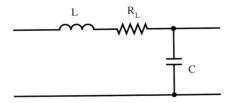

Fig. 7.13   *LCR* filter giving Butterworth or other responses.

Thus the Sallen and Key filter is equivalent to the LCR filter if we set:

$$L = 2C_2R^2 \qquad (7.58)$$

$$C = C_2 = C_1/2 \qquad (7.59)$$

$$R_L = 2R \qquad (7.60)$$

Note that the Sallen and Key filter is a composite structure incorporating strong negative feedback, and we cannot identify any *individual* set of components with the inductor. However, an inductor can be simulated using an operational amplifier by employing the circuit of Figure 7.14 (see also Section 2.10). It can also be shown that the impedance of this circuit is given by the following formula:

$$Z = [1 + \omega^2 C^2 R_1 R_2 + j\omega C(R_1 - R_2)]R_2/(1 + \omega^2 C^2 R_2^2) \qquad (7.61)$$

Clearly, this will only be inductive if $R_1 - R_2$ is positive. In addition, the frequency dependence will not be truly inductive unless $\omega^2 C^2 R_1 R_2$, $\omega^2 C^2 R_2^2 \ll 1$. If both of these conditions are valid, and also we make $R_1 \gg R_2$, then:

$$Z \approx R_2 + j\omega CR_1 R_2 = R_2 + j\omega L \qquad (7.62)$$

where

$$L \approx R_1 R_2 C \qquad (7.63)$$

If $R_1 = 1\,\text{M}\Omega$, $R_2 = 100\,\Omega$ and $C = 0.1\,\mu\text{F}$, we obtain an inductance $L = 10$ H – a remarkably high value for such convenient values of resistance and capacitance.

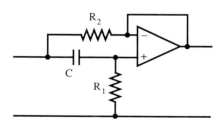

Fig. 7.14   Artificial inductor using an operational amplifier.

7.7 Summary

This chapter embarked on an analysis of two main operational amplifier circuits, both including negative feedback, obtaining simple formulae in each case – on the assumption of (a) infinite gain and (b) infinite input impedance. In practice these assumptions are sufficiently realistic and lead to useful conclusions being made about such circuits.

Detailed applications of these circuits were then made for low-pass, high-pass and band-pass filtering. However, the filter circuits initially examined (Sections 7.3–7.5) were first-order only, and in Section 7.6 better performance was shown to be obtainable with second-order filters of the Sallen and Key type. By varying the parameters in the latter type of circuit, a Butterworth "maximally flat" response, and also a Chebyshev response with limited pass-band ripple, was obtained. The equivalence of such circuits to LCR circuits was also considered; though it is often inconvenient to use inductors in practical circuits of these types, it is possible to simulate an inductor using an operational amplifier when this proves necessary. Higher-order filters can be constructed using these techniques: details are beyond the scope of the present volume.

7.8 Bibliography

Calvert and McCausland (1978) give a useful introduction to operational amplifiers, but do not discuss advanced filter design. Millman and Halkias (1972), Horowitz and Hill (1989) and Martin and Stephenson (1973) go into further depth, but Kuo (1962) should be referred to for fuller coverage of the subject. Basak (1991) gives a very readable account of the applications of operational amplifiers; see also Ahmed and Spreadbury (1973) for a more detailed but highly lucid treatment of operational amplifier circuits. An interesting viewpoint on active filters is provided by Girling and Good (1969); for the original paper describing the Sallen and Key filters see Sallen and Key (1955).

7.9 Problems

1. Give a full derivation of equation (7.12).
2. Generalize equations (7.8) and (7.12) to the case where the input impedance of the operational amplifier is Z_{in} rather than infinite.

3. Give full derivations of equations (7.43) and (7.47).
4. Find the responses, in decibels, of the following second-order filters at $\omega_c/2$, ω_c and $3\omega_c/2$: (a) a cascaded R-C filter, (b) a Butterworth filter, and (c) a Chebyshev filter with 0.5 dB ripple in the pass-band.
5. A fourth-order Butterworth filter is to be made using two Sallen and Key filter sections. Derive formulae for the filter from first principles (using equation 7.48), and determine the values of K that are needed for the two sections.
6. Give full derivations of equations (7.61–7.63).
7. Determine the voltage gain of the circuit shown in Figure 7.15, in which $R_1 = R(1-\eta)/2$ and $R_2 = R(1+\eta)/2$. Show that for a certain value of n the gain a_V varies between $-g$ and $+g$ as η is adjusted. For this value of n, show that a_V is a symmetrical function of η, with $a_V = 0$ for $\eta = 0$. Determine the maximum variation from linearity as a percentage of g.

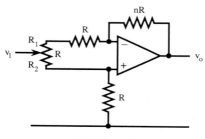

Fig. 7.15

8. The circuit in Figure 7.16 employs a 4-quadrant multiplier. Show that, as a result of feedback, the overall circuit acts as a *divider*. Devise a similar feedback circuit which acts as a *square-root* function generator.

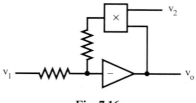

Fig. 7.16

8

Operational Amplifier Design

8.1 Introduction

In the diode circuit shown in Figure 8.1, it is well known that the voltage across the diode is ~ 0.6 V. In fact, this 0.6 V has a temperature coefficient of about -2.2 mV/$^\circ$C*.

Now consider the d.c. amplifier shown in Figure 8.2. This should have a voltage gain of at least 1000. However, the first transistor emitter-base junction acts exactly like the diode of Figure 8.1, and therefore causes a d.c. drift voltage of about -2.2 mV/$^\circ$C referred to the input of the amplifier: this means that at the output of the amplifier the drift voltage will be amplified to at least 2 V/$^\circ$C. Since temperature changes of $\pm 3^\circ$C are to be expected (especially in the vicinity of electronic equipment), the amplifier output is likely to swing wildly and indeed to spend much of its time completely saturated or cut off.

This highly unsatisfactory situation can only be overcome with the aid of fairly clever circuitry which is arranged to cancel drift voltages. The most simple arrangement for achieving this is the long-tail pair (Figure 8.3), which forms the basis for the work of this chapter.

---

* In fact, differentiation of the diode exponential law would suggest a figure of about $+2$ mV/$^\circ$C. However, leakage currents upset this prediction, and the overall drift has about the same magnitude but the opposite sign. Details of the theory underlying this effect are beyond the scope of this text.

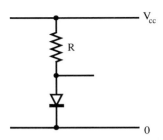

Fig. 8.1   Basic diode circuit resulting in significant thermal drift.

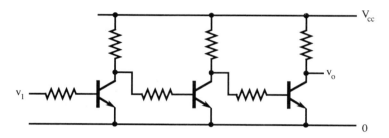

Fig. 8.2   Transistor d.c. amplifier subject to considerable thermal drift.

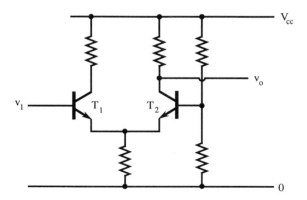

Fig. 8.3   Long-tail pair amplifier intended to minimize thermal drift.

8.2   The long-tail pair and its use in operational amplifiers

The action of the long-tail pair circuit is fairly straightforward: as T_1 (Figure 8.3) conducts more current, V_e rises, and T_2 conducts less current; likewise, as T_1 conducts less current, T_2 conducts more, so a sort of see-saw action occurs. We shall analyse the action of the long-tail pair in more detail below. Meanwhile, note that the long-tail pair circuit shown in Figure 8.4 employs a constant current source in the tail; this means that as T_1 conducts more current, T_2 conducts an *equal* amount less current, so the collector voltages are ideally forced to go up and down by equal amounts on application of an input voltage to the base of T_1. The important thing to notice about such circuits is that the drifts of the two transistors tend to cancel, if the output signal is taken as the *difference* in voltage at the two collectors. Indeed, if the two transistors are on the same substrate and are identically constructed, then their drift voltages normally cancel to within $\sim 0.1\%$ over a considerable range of temperature.

As noted earlier, the long-tail pair is extensively used in operational amplifier design, almost always in its ideal form incorporating a constant current tail. Figure 8.5 shows the approximate equivalent circuit of an early operational amplifier, the 702. Note the two long-tail pairs, feeding the output via two emitter followers. The first of these is used to feed a resistor which drops a fixed voltage, this being achieved with the aid of a constant current source. Note that the constant current sources in the tails of the two long-tail pairs are constructed using current mirrors, to give good current stability within the constraints of a limited overall supply voltage.

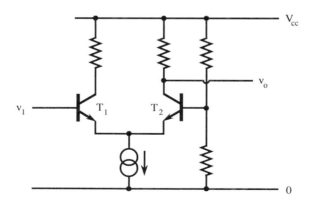

Fig. 8.4   Ideal form of long-tail pair using constant current source.

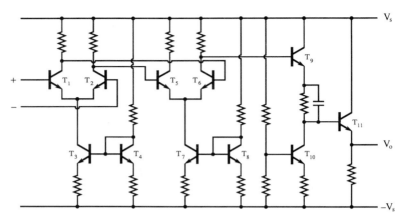

Fig. 8.5 Basic design model of 702 operational amplifier.

The 709 is similar to the 702, but its second long-tail pair uses Darlington pair transistors, and this pushes up the overall voltage gain to ∼ 50,000. The 741, which is perhaps the most widely used operational amplifier, has a voltage gain ∼ 200,000. This large value is attained by including a long-tail pair whose load resistors are replaced by a current mirror: this has the effect of (a) performing differential to single-ended conversion, so the current gain produced by the first transistor is not wasted and overall gain is doubled, and (b) feeding the collector of the second transistor into an "active" high impedance (constant current) load, thereby significantly increasing the voltage gain (see Section 3.3). A Darlington pair amplifier is also used. Figure 8.6 shows a *very* simplified version of the overall circuit. A further interesting feature is the use of a V_{be} multiplier active voltage drop circuit, fed from a constant current source, in place of a resistor to bias the output transistors. Finally, the capacitor provides a dominant time-constant, which ensures stability against oscillation when the operational amplifier is included in feedback networks (see Section 5.8).

Overall, operational amplifier design incorporates a number of interesting circuit configurations and features which involve vital fundamental principles. These circuit features may be summed up as follows:

1. long-tail pair differential amplifiers with constant current tail drivers;
2. current mirror constant current sources;
3. differential to single-ended conversion;
4. active high-impedance loads;
5. Darlington and complementary Darlington high current gain pairs;

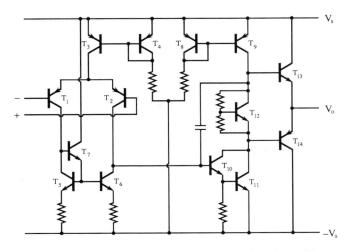

Fig. 8.6 Basic design model of 741 operational amplifier.

6. push-pull circuit for high current output on rising and falling voltage swings;
7. V_{be} multiplier with constant current source;
8. single capacitor dominant time-constant.

8.3   Analysis of the long-tail pair configuration

First, we calculate the differential gain of the long-tail pair. Consider a perfectly balanced long-tail pair (Figure 8.7) and let v_1 increase and v_2 decrease by equal amounts. The resulting changes in base currents will not themselves affect v_e. Clearly, the changes in the collector currents will be equal and opposite, and again will not affect v_e. Hence we can proceed assuming that v_e stays constant. We now have:

$$v_{c1} = -R_c i_{c1} = -R_c h_{fe} i_{b1} = -R_c h_{fe}(v_1 - v_e)/h_{ie} \qquad (8.1)$$

$$v_{c2} = -R_c i_{c2} = -R_c h_{fe} i_{b2} = -R_c h_{fe}(v_2 - v_e)/h_{ie} \qquad (8.2)$$

$$\therefore \quad v_{c1} - v_{c2} = -R_c h_{fe}(v_1 - v_2)/h_{ie} \qquad (8.3)$$

$\therefore$ differential gain,

$$a_{\text{diff}} = -R_c\left(\frac{h_{fe}}{h_{ie}}\right) = -R_c y_{fe} \qquad (8.4)$$

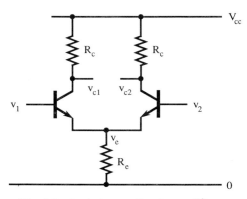

Fig. 8.7 Basic long-tail pair amplifier.

Next, we calculate the "common mode" gain of the long-tail pair. In this case we start with the circuit perfectly balanced and keep it balanced. Since both halves of the circuit operate identically, it is permissible to split the circuit into two identical parts, and just analyse one of them (Figure 8.8). (Note that $2R_e \| 2R_e = R_e$.) We now have:

$$v_{c1} = -R_c i_{c1} = -R_c h_{fe} i_{b1} \tag{8.5}$$

$$v_e = 2R_e i_{e1} = 2R_e (h_{fe} + 1) i_{b1} \tag{8.6}$$

$$v_1 - v_e = h_{ie} i_{b1} \tag{8.7}$$

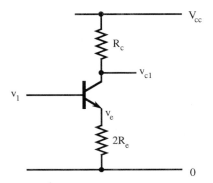

Fig. 8.8 One half of long-tail pair with symmetrical configuration of voltages. This *half* long-tail pair is useful for calculating common mode gain.

Eliminating v_e, we get:

$$v_1 = [h_{ie} + 2R_e(h_{fe} + 1)]i_{b1} \tag{8.8}$$

and eliminating i_{b1}, we get:

$$\frac{v_{c1}}{v_1} = \frac{-R_c h_{fe}}{h_{ie} + 2R_e(h_{fe} + 1)} \tag{8.9}$$

This is the common mode gain a_{cm}. Since $h_{fe} \gg 1$, it is usually adequate to approximate this quantity in the form:

$$a_{cm} = -R_c/2R_e \tag{8.10}$$

Since $y_{fe}^{-1} \ll R_e$ (typical values being $4\,\Omega$ and $500\,\Omega$ respectively), it is clear that $a_{diff} \gg a_{cm}$. Thus the common mode rejection ratio (CMRR) a_{diff}/a_{cm} is quite large:

$$\text{CMRR} = a_{diff}/a_{cm} = 2R_e y_{fe} \approx 2 \times 500/4 = 250 \tag{8.11}$$

On the other hand, if the long-tail pair included a simple constant current source of output impedance $3\,\text{k}\Omega$ (which should easily be available using a BC107 or other silicon transistor), the CMRR would rise to a phenomenal 1500!

Finally, note that matched BJTs in close proximity on the same substrate can now be made with V_{be} values matching within $\sim 50\,\mu\text{V}$, V_{be} drift values matching within $1\,\mu\text{V}/°\text{C}$, and h_{fe} values matching within $\sim 1\%$. These figures imply that a normal ambient temperature variation of 10 degrees produces a temperature difference at the transistors of less than 10 millidegrees! It is left to the reader to estimate how these figures affect the CMRR values derived above.

8.4 Summary

This chapter proceeded to discuss d.c. amplifier design and showed that temperature-induced drift is a severely limiting factor at moderate to high gains. The need to cancel thermal drift led naturally to the intrinsic value of the long-tail pair amplifier as a component circuit. The differential and common mode gains of this type of amplifier, and its common mode rejection ratio (CMRR), were therefore calculated. However, a number of other circuit configurations and tricks* (some of them already described in earlier chapters) are commonly used in operational amplifiers: these were enumerated, and some *much* simplified models of early operational amplifiers were described. In many ways the work of this chapter is a revision of techniques discussed earlier, and thereby completes the basic electronics covered in this book. In this light the work of the next chapter provides further rehearsal of basic ideas, while it moves into a vital area of application.

8.5 Bibliography

Long-tail pairs are covered well by Calvert and McCausland (1978). For further detail on operational amplifier design see Watson (1989) and Basak (1991).

8.6 Problems

1. Show that if the circuit of Figure 8.1 contains an *ideal* diode obeying the exponential law, then the voltage across it will drift by $+2\,\mathrm{mV/^{\circ}C}$, as stated in Section 8.1.

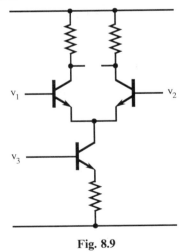

Fig. 8.9

---

*It is perhaps misleading to refer to these as circuit tricks: voltage level-shifting circuits, Darlington pairs, current mirror circuits, complementary pairs and so on are hardly non-rigorous "trick" circuits – the trick lies in combining these circuits in such subtle ways that the bystander is initially surprised how hard his understanding is being pressed by the final design!

2. Show that a long-tail pair can be used as a multiplier if its tail resistor is replaced by a 1-transistor voltage-to-current converter, as in Figure 8.9. Determine whether the resulting circuit is a 1-, 2- or 4-quadrant multiplier (see also Problem 2.5).

3. Generalize the formulae of Section 8.3 to the case of imperfectly matched long-tail pair transistors – e.g. with different values of h_{fe}. Use the values given at the end of Section 8.3 to derive typical values of a_{cm}, a_{diff} and CMRR for a long-tail pair.

4. The circuit of Figure 8.5 is found to have an overall gain of 20,000. Estimate the thermal drift at its output and determine whether there is any risk of the circuit becoming saturated or cut off if the ambient temperature rises by 10 °C on a hot day. Assume that the circuit is well balanced initially.

9

Stabilized Power Supplies

9.1 Introduction

When electronic circuits are designed, it is normally assumed that a power supply will be available that is capable of providing certain d.c. voltages at suitable current levels, and it is perhaps implicitly assumed that the d.c. voltages will be completely stable under all circumstances. Of course, it is impossible to fulfil this assumption exactly in practice. However, it can be achieved to most desired levels, with sufficient design effort. In this chapter we consider the problems of designing stabilized power supplies, and in particular we examine their specification.

9.2 Evolving a strategy for stabilized power supply design

In many cases, the power for a circuit will be taken from the a.c. mains, and will then be transformed down to some convenient voltage and full-wave rectified (Figure 9.1). This leads to the task of smoothing the full-wave rectified waveform and stabilizing the voltage at some desired value. Next, note that the output voltage will have to be stable against changes in mains voltage and against changes in its output current; furthermore, the full-wave rectified waveform will have to be suppressed, giving insignificant levels of output ripple; in addition, negligible noise and thermally induced or other drift should remain. There will also be other requirements such as suppression of mains-borne voltage spikes, output short-circuit protection, and perhaps operation over a range of output voltages, with some convenient means of adjustment.

(a)

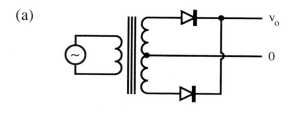

(b)

Fig. 9.1 Full-wave rectification. The circuit in (a) performs full-wave rectification of the transformed a.c. supply, with the result shown in (b).

Thus, the most obvious specification parameters for a power supply are:

1. mains supply voltage;
2. output stabilized d.c. voltage, or range of output voltages;
3. output current range;
4. current at which short-circuit protection switches on;
5. temperature coefficient of output voltage, often called "drift";
6. remanent ripple v_r and noise v_n levels;
7. output impedance, Z_{out}, which gives an indication of how much V_{out} drops as more current is drawn from the power supply;
8. stability factor, η, defined as the fractional change in output voltage divided by the fractional change in mains voltage: $\dfrac{\mathrm{d}V_{out}}{\mathrm{d}V_{mains}} \Big/ \dfrac{V_{out}}{V_{mains}}$.

In respect of the ripple voltage v_r, which is the remanent a.c. voltage appearing at the output, note that for *full*-wave rectified mains, the ripple frequency is doubled, to 100 Hz (which incidentally makes it easier to smooth). Actual figures for the output ripple frequently include any small noise voltages, giving a total ripple $\sim 1\,\mathrm{mV}$.

The simplest type of voltage stabilizer (if we ignore any preliminary R-C smoothing filter) is probably a high-power Zener diode and resistor combination (Figure 9.2), which might appear capable of reasonable performance if the dynamic resistance of the Zener diode is less than $1\,\Omega$. However, such a circuit is inadequate for powering (say) an audio amplifier feeding a loudspeaker. We can improve its performance significantly by using

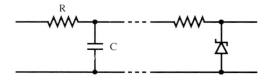

Fig. 9.2 A simple form of voltage stabilizer. The smoothing circuit on the left is fed from a full-wave rectified supply (see Figure 9.1), and then applied to the resistor–Zener diode combination on the right. This circuit performs a rudimentary stabilization function, but is inefficient since the Zener diode consumes a large proportion of the available power, while yielding a rather low stability factor. The arrangement is little used in practice.

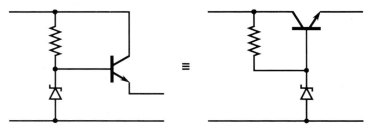

Fig. 9.3 A basic but effective form of voltage stabilizer. This circuit uses a Zener diode to provide a suitable reference voltage, and a large current is provided at (approximately) this voltage by an emitter follower. Note the two ways such circuits are commonly drawn.

a high-stability (low-power) Zener diode to feed an emitter follower (Figure 9.3). Ultimately such a "feed-forward" design is limited, as it works by calculating what voltage to put at the output, rather than by sensing what it is and correcting it*. Thus the better designs include powerful output voltage sensing arrangements and negative feedback loops which are designed to stabilize them. In that case some fraction of the output voltage is fed to a difference amplifier whose other input is connected to a stabilized voltage source (normally a Zener diode).

---

* In fact, this is not entirely valid, as an emitter follower incorporates local negative feedback, and *is* sensitive to what is happening at the output. However, the crucial point is that it is better to include an *explicit* negative feedback loop which can rigorously stabilize the output voltage. This is particularly important if a power supply is used to stabilize a remote voltage, where the cable carrying the current is subject to a significant voltage drop: in that case provision has to be made for use of long voltage sensing leads.

In Figure 9.4 the difference amplifier is shown as an operational amplifier, though a simple design using a discrete component long-tail pair can be quite effective. Note that the operational amplifier works at some base voltage level (such as that of a Zener diode) that is often well below the output voltage level. This means that the d.c. level of the output voltage has to be lowered artificially: in the circuit of Figure 9.4 this is achieved using a resistor network. However, this method also attenuates the output voltage variation, and hence reduces the gain of the feedback loop, while this in turn reduces the power supply stability factor. For this reason the output voltage may better be communicated to the differential amplifier via a voltage level-changing Zener diode.

Thus the final type of stabilizer circuit described above incorporates a negative feedback loop of loop gain $\lambda < 0$, where $|\lambda| \gg 1$. The result of feedback is to reduce Z_{out}, η and v_r, *all* by a factor $\sim |\lambda|$, and it is clearly the value of λ that decides many of the vital properties of this type of circuit.

Stabilized power supplies are of such vital importance that much effort has gone into their design and implementation. It is nowadays rarely necessary to design a power supply of the type mentioned above because special integrated circuits are available that incorporate all relevant components. However, in high-current applications it is usual to employ an integrated circuit that is designed so that a single high-current output transistor can be added for the specific purpose in hand.

Finally, we mention the switched mode type of power supply that is now quite common. This uses a stream of narrow current pulses to feed a smoothing circuit, and the number or duration of the pulses is carefully controlled by a feedback loop in such a way as to maintain the output d.c.

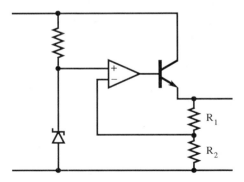

Fig. 9.4 A more practical form of voltage stabilizer. This circuit is similar to the basic circuit of Figure 9.3, but uses an operational amplifier to increase the negative feedback loop gain and thereby increase the stability factor.

voltage at the required level. Design of such circuits is a rather specialized topic and will not be covered in detail here, but their main attraction is their compactness – resulting from the fact that switched devices are used; since these devices are always on or off, they consume little power and therefore the circuit can be made efficient and highly compact.

9.3 Characteristics of Zener diodes

It is perhaps obvious that the voltage stability of the stabilized power supplies described above will depend markedly on the quality of the reference voltage source they employ. So far it has been assumed that Zener diodes will be used for this purpose. In fact Zener diodes are not automatically the best choice, and much depends on their characteristics. Relevant characteristics include:

1. the dynamic resistance;
2. the thermal stability;
3. the r.m.s. noise voltage that is generated internally.

   In fact, the dynamic resistance varies markedly with voltage for any given series of Zener diodes, and is typically a minimum at about 6 V (see Figure 9.5) – while it is so high below ~2.7 V that Zener diodes are not manufactured in this range. It also happens that at about 6 V, Zener diodes change from a positive to a negative temperature coefficient of voltage (Figure 9.6): the reason for this is that the true Zener (quantum mechanical tunnelling) effect, which is dominant below ~6 V, gives way above this voltage to the avalanche effect, wherein accelerating charge carriers gain enough energy to generate more electron-hole pairs in the semiconducting material. From an engineering

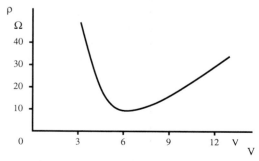

Fig. 9.5 Variation of dynamic resistance for a typical series of Zener diodes. Notice that the dynamic resistance ρ is a minimum for Zener diodes with a voltage ~6 V.

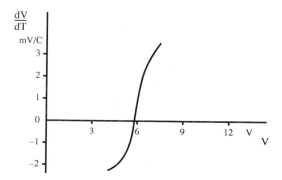

Fig. 9.6 Variation of temperature coefficient for a typical series of Zener diodes. Notice that the temperature coefficient of voltage is zero for Zener diodes with a voltage ~6 V.

point of view it is therefore best to use Zener diodes of around 6 V: in that case a resistive attenuator may have to be used to reduce *either* the Zener voltage *or* the supply output voltage to similar values so that the feedback loop can stabilize their difference at zero. Returning to the dynamic resistance of the Zener diode, and the loss of voltage stability that this can cause, note that stabilization of the current supply feeding the Zener can help its voltage stability considerably by specifying the exact working point on the *I-V* characteristic. To achieve this improvement, nothing more than a single (2-terminal) FET constant current source device may be required!

Finally, we consider the noise generated by Zener diode voltage reference sources. Although the Zener effect and the avalanche effect might be expected to generate considerable statistical fluctuations in the output voltage, in practice this is not as serious a problem as the noise that arises from surface effects in the semiconducting material. However, "buried" (subsurface) Zeners have been designed which exhibit remarkably low output noise levels, in the microvolt range. More generally, it should be noted that careful selection of individual electronic components (not just Zener diodes) will often minimize noise levels – the engineer should always be aware of the possibility of "rogue" components that are considerably more noisy than the norm.

9.4 Bandgap reference

Because of the especial importance of high-stability voltage sources, a great amount of effort has recently been devoted to developing a new type of voltage reference, called the bandgap reference. The strategy adopted is to try to eliminate the temperature dependence of the forward voltage drop V_{be} of the emitter-base diode in a junction transistor. Experimentally, it is found that V_{be} has a temperature coefficient of about $-2.2\,\mathrm{mV/°C}$. However, if we differentiate the exponential law for the diode, we can predict that V_{be} should have a temperature coefficient of about $+2\,\mathrm{mV/°C}$. This discrepancy can be explained in terms of the strong variation of leakage current with temperature. (It arises since the actual leakage current in the emitter-base diode consists of the ideal junction leakage current of about $10^{-15}\,\mathrm{A}$ plus a much larger current of about $10^{-8}\,\mathrm{A}$ which is due to additional currents flowing near the surface of the semiconductor.)

In the bandgap reference, an attempt is made to cancel out these two temperature variations. Fortunately, although the actual leakage current is quite large relative to the ideal junction leakage current, it does not contribute to the transistor action. Hence a *differential* effect, with a positive temperature coefficient of $2\,\mathrm{mV/°C}$, can be induced between two nearly identical transistors. This differential effect ΔV_{be} can be obtained by means of a current mirror in which one transistor has been unbalanced by means of an emitter resistor R_e, and then ΔV_{be} appears across this resistor (Figure 9.7). This value of ΔV_{be} is amplified to $\Delta V_{be} \times (R_c/R_e)$ in a collector load R_c at the output of the current mirror, and this voltage is added to the V_{be} of a third transistor. Adjustment of R_c/R_e then ensures correct cancellation of the temperature coefficient. In fact, it is obvious from the above figures that this will occur when $\Delta V_{be} \times (R_c/R_e) \approx V_{be} \approx 0.6\,\mathrm{V}$, so the output voltage of the device will be about $1.2\,\mathrm{V}$. It turns out that exact cancellation occurs when

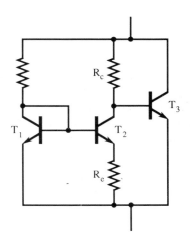

Fig. 9.7   Bandgap reference circuit.

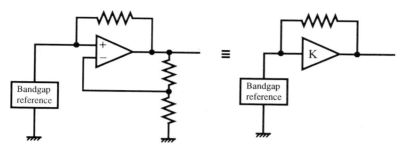

Fig. 9.8   Circuit for obtaining desired voltage from a bandgap reference.

the voltage is equal to the silicon bandgap voltage extrapolated to absolute zero, namely 1.205 V, though for room temperature devices the correct figure is ~1.235 V. Note that by a rather subtle trick, the current mirror is provided with a constant current using the very voltage it has helped to stabilize!

Such devices are marketed* by a number of manufacturers, and have temperature coefficients measured in parts per million per degree C (i.e. in this case ~1 μV/°C). It should also be noted that they operate in the crucial region below 3 V where normal Zener diodes are unavailable. On the other hand, a bandgap reference voltage can conveniently be magnified to any normal level by an operational amplifier using the circuit of Figure 9.8, while at the same time feeding it with a constant current for extra stability.

9.5   Protection circuits

There are several applications in which an emitter follower or similar buffer stage feeds an unknown load. One is at the output of an operational amplifier, and another is in a stabilized voltage power supply. In such cases it is common practice to include a safety circuit which limits the output current automatically (in case the output is inadvertently short-circuited).

In its simplest form, the extra current-limiting circuitry consists of a transistor plus a low-value resistor, arranged so that the transistor is normally off, but switches on as soon as the voltage across the resistor rises excessively. The collector of the transistor is connected to the base of the emitter follower transistor, and prevents it from delivering too high a current. The modified emitter follower circuit is as shown in Figure 9.9. If the maximum permitted

---

* Needless to say, such circuits have to be fabricated on a single i.c. substrate, so that all components are accurately matched and maintained at the same temperature.

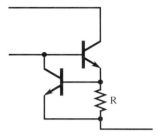

Fig. 9.9 Short-circuit protection device. This circuit replaces the emitter follower in more basic stabilized power supply circuits, such as those of Figures 9.3 and 9.4. This circuit has the characteristic of reverting to normal operation when the current overload is removed.

current is to be I_{max}, then R should be such as to produce a voltage drop $\sim 0.6\,$V for $I = I_{max}$.

The circuit described above provides a very simple form of short-circuit protection which basically maintains the output current at I_{max} as the load resistance decreases. Since it does not actually remove the excessive current, the alternative "foldback" approach was devised: this reduces the output current linearly with decrease of output voltage V_{out}, once the output current has exceeded I_{max}. However, it is necessary for the power supply stabilization loop to be self-starting, and therefore it is normal for the design value of the short-circuit output current to be set at $\sim I_{max}/3$ (i.e. the output current derates from I_{max} at normal output voltage to $I_{max}/3$ at $V_{out} = 0$).

In many applications it is necessary to protect a circuit from voltages greater than a certain limiting value, in case of failure of the power supply (and especially in case the output transistor goes short-circuit). For example, TTL logic circuits are likely to fail if their 5 V power supply exceeds 7 V, even momentarily. Thus over-voltage protection circuits are useful. These can consist of an SCR "crowbar" which is used to short the output of the power supply. In this case, the SCR is triggered to switch on when the output voltage exceeds the specified maximum. Note that SCRs only stop conducting when the current is interrupted, e.g. by removing the mains supply from the circuit.

9.6 Summary

This chapter starts with a full-wave rectified a.c. waveform, considers coarse smoothing of it, and then proceeds to examine how a fully stabilized d.c.

voltage supply can be obtained from it. The obvious approach of employing an emitter follower fed from a Zener reference voltage gives reasonable results. However, for high performance an explicit feedback loop is required, and this is most effective if it incorporates a high-gain operational amplifier.

As an adjunct to this approach, some attention to the reference voltage is necessary, since the stability of the output cannot be better than that of the reference. A relatively recent device called the bandgap reference is described, and the full circuit for this incorporates a number of subtleties. The ultimate accuracy of this circuit can only be quoted in this chapter – a full analysis requires calculations based on models of the energy-band structure of the semiconducting material and is beyond the scope of this volume.

9.7   Bibliography

Probably the best and most comprehensive reference on this topic is Horowitz and Hill (1989). Not only does this book provide a sound exposition of the various types of stabilized power supply (including switched mode supplies), but it enunciates the faults they are subject to. As usual, it has a wealth of detail on relevant devices, including a *very* full catalogue of the various types of Zener diode, with their temperature coefficients, noise characteristics, and so on. For the theory of bandgap references, see for example, Widlar (1971), Rehman (1980) and Watson (1989).

9.8   Problems

1. For the circuit of Figure 9.4, determine whether the remanent a.c. voltage across the output transistor or that across the Zener diode has the greater effect on the output ripple; assume the BJT parameters of Chapter 1 and the Zener parameters of Chapter 9.
2. Short-circuit protection on a stabilized voltage power supply can be either constant current (stabilized at some maximum permitted current) or "foldback" (current derated linearly as the output voltage falls). Determine which of these situations, if either, occurs if the modification shown in Figure 9.9 is applied to (a) the circuit of Figure 9.3, and (b) the circuit of Figure 9.4.
3. Sketch the output voltage characteristic of a stabilized voltage power supply with short-circuit protection, which gives (a) a constant current

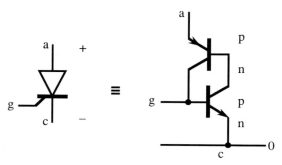

Fig. 9.10

or (b) foldback. Superimpose on your diagram the load lines for various load resistances, and determine the conditions under which each type of short-circuit protection has a single working point so that it switches on automatically.

4. An SCR is a pnpn device which can be thought of as two complementary transistors in the configuration of Figure 9.10, with three connections, cathode c, anode a, and gate g. Show that if the gate terminal g is kept at 0 V relative to the cathode, then there will be no anode current, but that once g exceeds 0.6 V the SCR will become switched on; the device will then conduct until the anode supply is interrupted. Devise a circuit for providing over-voltage protection using an SCR.

Part 2
Noise

This short section of the book introduces the important topic of noise in electronic systems. Chapter 10 covers the origins and basic properties of noise, while Chapter 11 concentrates on its appearance in amplifiers, and means by which it may be measured and minimized. Although this part of the book considers how noise may be reduced, special noise-limiting filters are not considered here, since they are dealt with in depth in Part 3 under the heading *signal recovery*.

10

Noise and Its Origins

10.1 Introduction

As we have seen in the foregoing chapters, much of analogue electronics is concerned with the design of devices such as amplifiers, oscillators and power supplies: this is generally a fairly straightforward process in the sense that a specification is made about the types of signal and how they are to be processed, and the design work then proceeds accordingly. However, if we are dealing with very small signals, we have to take account of various forms of electrical noise and interference, some of which can make the lives of electronics design engineers rather difficult. In particular, we have to deal with the following:

1. Thermally generated or "Johnson" noise, which is ultimately due to the random thermal motions of the electrically charged carriers that exist in any electronic circuit.
2. "Shot" noise, which is due to the discrete nature of charge carriers, and arises since any charge transfer involves an integral number of charge carriers.
3. "Flicker" noise, which is due to rather odd effects on the surfaces of solid state components in electronic circuits, with the result that the noise contributions are greatest at very low frequencies (a $1/f$ law is generally adhered to quite closely).
4. Remanent a.c., typically at 50 Hz or 100 Hz, that has not been eliminated by the stabilized power supply feeding the circuit.
5. Pickup from nearby mains lines, again at 50 Hz or 100 Hz, which is ultimately due to poor circuit layout or inadequate screening.
6. Radio-frequency interference from airborne signals such as lightning, radar or radio transmitters, or even domestic or other appliances (e.g. light

Fig. 10.1   Typical noise waveform.

dimmers, microwave ovens, including their microwave generators and internal microprocessor clocks). Again, these problems tend to arise as a result of inadequate screening.

7. Mains-borne interference from lightning and switching transients, which can cause sharp impulses in mains voltage that may pass right through a stabilized power supply (on the whole the latter are designed to suppress low frequencies, but may let through fast transients).

8. "Crosstalk", which is the pickup of unwanted signals from lines that lie across the path carrying the currently relevant signal.

9. "Clutter", which is the presence of unwanted signals on the line carrying the currently relevant signal. (Clutter may arise as a result of crosstalk, but more often it refers to unwanted signals that are picked up by the same sensor that detects the desired signals – a situation that is of great importance, for example, with radar receivers.)

The first three of these sources of noise are unstructured and appear on an oscilloscope screen or pen recorder chart (on which the gain control has been turned up sufficiently) as in Figure 10.1. The remaining sources listed above tend to have particular frequency content, or else (as for interference from lightning) to have a specific rather short time-span: hence in either of these cases the interference is highly structured and easily recognizable. We shall consider in some detail the case of thermally generated noise, since it is generally the most important source of noise in electronic circuits.

10.2 Thermal noise

Surprising as it may seem, the simple resistor is a major contributor of noise to electronic circuits. A full treatment of this topic is beyond the scope of the present discussion. However, we can get some insight into the nature of

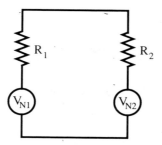

Fig. 10.2   Circuit used for exploring resistor noise.

thermal noise (also known as *Johnson noise or resistance noise*) by considering two resistors connected together, as in Figure 10.2.

First, we know that a resistor cannot generate a mean voltage that is different from zero: otherwise a steady measurable current would flow around a circuit such as that in Figure 10.2. Suppose instead that a resistor R generates a mean square noise voltage proportional to $g(R)$. If the instantaneous noise voltage generated in R_1 is V_{N1}, this produces a current $V_{N1}/(R_1+R_2)$, and the instantaneous power dissipated in R_2 is $R_2 \times V_{N1}{}^2/(R_1+R_2)^2$. Similarly, if the instantaneous noise voltage generated in R_2 is V_{N2}, the instantaneous power dissipated in R_1 is $R_1 \times V_{N2}{}^2/(R_1+R_2)^2$. But there can be no net mean power transfer from R_1 to R_2, or vice versa, if both resistors are to remain at the same temperature. Therefore we have:

$$g(R_1)R_2/(R_1+R_2)^2 = g(R_2)R_1/(R_1+R_2)^2 \qquad (10.1)$$

so

$$g(R_1)/R_1 = g(R_2)/R_2 \qquad (10.2)$$

which must be a constant independent of the value of resistance. The only way this can be achieved is if $g(R) \propto R$. In fact the full Johnson noise formula is:

$$\overline{V_N{}^2} = g(R) = 4kTRB \qquad (10.3)$$

where B is the bandwidth of the observing system (e.g. $B=20$ MHz for a 20 MHz oscilloscope), and k is Boltzmann's constant (1.38×10^{-23} J/K). Since frequency does not appear explicitly in this formula, resistor noise can be presumed to have a flat response: this type of noise is commonly called "white", by analogy with light, in that the response is the same over a wide range of frequencies. It is important to note that this characteristic, and the mean square noise voltage given by the above formula, applies to an *ideal* resistor. Real resistors generate at least this ideal amount of noise, but they may produce even more noise if they have specific faults*. Finally, it should be noted that it is not just resistors that generate noise, but any circuit component which has an equivalent circuit containing a resistance – e.g. a diode, transistor or other device. However, a *pure* inductance or capacitance does not generate noise: indeed, such components usually operate in such a way as to cut down bandwidth, and therefore have the effect of *reducing* the observed noise for a given circuit.

Finally, we consider what happens when noise is generated by a resistance R_1 and is matched to a load resistance R_2, so that maximum power transfer occurs. In that case $R_1 = R_2$ and the noise power generated by R_1 and dissipated in R_2 is:

$$P_N = R_1 \times \overline{V_{N1}^2}/(R_1 + R_1)^2 = V_{N1}^2/4R_1 = kTB \tag{10.4}$$

This quantity is called the *available noise power*.

10.3 Shot noise

Shot noise is a further important source of noise that occurs in electronic circuits. It arises because electric currents are carried by electrons which have a discrete charge, so that fluctuations in the numbers of charge carriers passing a given point give rise to random additive noise currents. Schottky showed that this effect leads to a distribution of current values about the mean current I, and is given by the mean square current:

$$\overline{i_n^2} = 2eIB \tag{10.5}$$

e being the electronic charge, and B being the bandwidth. *A priori*, shot noise is white and thus has infinite bandwidth, but as soon as this is restricted

---

* In fact, it is not possible for a pure resistor to produce more noise than the ideal amount without contravening the second law of thermodynamics. However, a faulty (e.g. cracked) resistor could be an ideal resistor in parallel with a small capacitor plus resistor. Then it would generate more noise at high frequencies than its d.c. resistance would indicate. In addition, any piezo-electric impurities might give unpredictable noise voltages. Finally, a cracked resistor, or one with faulty connections, might alternate under physical strain between a set of resistance values, thereby leading to variations in output voltage: such effects are not at all uncommon in discrete-component electronic circuits.

by stray capacitance, etc., to some finite value B, the noise becomes limited and its magnitude may be estimated from the above formula. In fact the latter merely indicates the mean square value of the noise amplitude distribution, and this distribution turns out to be a Gaussian probability function (see below).

It is important to realize that shot noise is usually associated with devices not in thermal equilibrium. Examples are a thermionic valve in which electrons emitted by the cathode have been accelerated towards the anode, and a semiconductor diode or transistor in which minority carriers have climbed a potential barrier. However, in the latter case, if there is no externally applied potential so that thermal equilibrium exists, the shot noise formula leads to identically the same result as the Johnson noise formula: details of the calculation are presented below, in Section 10.7.

Finally, it is worth noting that once shot noise has been generated by varying numbers of carriers passing a given point, it may be amplified by subsequent parts of a circuit (as in a transistor), or smoothed (as in a space-charge limited thermionic valve): thus it may *appear* that equation (10.5) is not obeyed in practical situations.

10.4   Gaussian noise and spectral whiteness

Gaussian noise is noise with a Gaussian amplitude probability distribution (Figure 10.3). In that case the probability density distribution of amplitudes of (say) voltage is given by:

$$p(V) = \frac{1}{\sqrt{2\pi\sigma^2}} \exp[-(V - \bar{V})^2/2\sigma^2] \qquad (10.6)$$

(a)   (b)

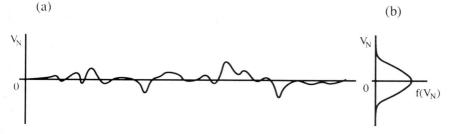

Fig. 10.3   Gaussian voltage distribution. Here (a) shows a Gaussian noise waveform, and (b) shows the distribution of voltages for the waveform.

where σ^2 is the mean square variation of the measured voltages:

$$\sigma^2 = \overline{(V - \overline{V})^2} \tag{10.7}$$

It is important to note that Gaussian noise is not necessarily white noise, and white noise is not necessarily Gaussian. In fact, unconstrained Johnson noise is both Gaussian and white, and the same comment applies for shot noise: by definition, any form of narrow-banding prevents these forms of noise from being white, but need not affect the Gaussian nature of the noise.

10.5 Combining noise voltages

In Section 10.1 we listed nine types of noise and interference. Of these only the first three are completely random, the last six being structured – e.g. being at some supply or pickup frequency or having the time-profile of some interfering signal. To determine the overall effects of random noise voltages from separate sources all we have to do is to add the mean square voltage variations:

$$\overline{V_N^2} = \sum_i \overline{V_{Ni}^2} \tag{10.8}$$

When the individual noise components have Gaussian voltage distributions, the mean square voltages are equal to the variances of the various distributions, and the final distribution is also Gaussian. When the individual distributions are not Gaussian, the final distribution will no longer be Gaussian, but generally the overall variance will still be equal to the sums of the variances, so adding the mean square voltages is still meaningful. Ultimately, the reason that this is so arises from the definition of the variance of a distribution. For suppose $V_N = V_{N1} + V_{N2}$. Then we can calculate the variance of V_N as follows:

$$\sigma_N^2 = \overline{(V_N - \overline{V_N})^2}$$

$$= \overline{[(V_{N1} + V_{N2}) - \overline{(V_{N1} + V_{N2})}]^2}$$

$$= \overline{[(V_{N1} - \overline{V_{N1}}) + (V_{N2} - \overline{V_{N2}})]^2}$$

$$= \overline{(V_{N1} - \overline{V_{N1}})^2} + 2\overline{(V_{N1} - \overline{V_{N1}})(V_{N2} - \overline{V_{N2}})} + \overline{(V_{N2} - \overline{V_{N2}})^2}$$

$$= \sigma_{N1}^2 + 2C_{N12} + \sigma_{N2}^2 \tag{10.9}$$

where we have assumed that

$$\overline{(V_{N1} + V_{N2})} = \overline{V_{N1}} + \overline{V_{N2}} \tag{10.10}$$

Now the first and the third terms in equation (10.9) are the variances of the individual noise voltage distributions, while the second term is due to the *covariance* C_{N12} between the two noise distributions. The latter quantity will approximate to zero in many practical situations, since $(V_{N1} - \overline{V_{N1}})$ will be as often positive as negative for each value of $(V_{N2} - \overline{V_{N2}})$, and vice versa. In that case we have:

$$\sigma_N^2 \approx \sigma_{N1}^2 + \sigma_{N2}^2 \tag{10.11}$$

Many noise distributions can be taken to be symmetric – as in the case of the Gaussian. However, the above calculation covers situations where the individual distributions may not be symmetric. Thus it applies even for the case of interfering signals (structured noise) which are not symmetric, but only so long as they are completely uncorrelated with each other: otherwise the covariance term from equation (10.9) will have to be retained*.

10.6   Degree of correlation in the time domain

An important property of noise is its degree of correlation in the time domain. Now the autocorrelation function $R(\tau)$ of a signal $v(t)$ is closely related to the Fourier transform of its power spectrum $P(f)$, via the formulae:

$$R(\tau) = \lim_{T \to \infty} \frac{1}{T} \int_{-T/2}^{T/2} v(t)v(t+\tau)\,dt \tag{10.12}$$

$$R(\tau) = \int_{-\infty}^{\infty} P(f)\exp(j\omega\tau)\,df \tag{10.13}$$

where

$$P(f) = \int_{-\infty}^{\infty} R(\tau)\exp(-j\omega\tau)\,d\tau \tag{10.14}$$

---

*An obvious case in which this happens is when the two noise voltages ultimately have the same physical origin, so that $V_{N2} = \gamma V_{N1}$. In that case $\sigma_N^2 = (1+\gamma)^2\sigma_{N1}^2$ rather than $(1+\gamma^2)\sigma_{N1}^2$, e.g. if $\gamma = 1$, $\sigma_N^2 = 4\sigma_{N1}^2$ rather than $2\sigma_{N1}^2$, while if $\gamma = -1$, $\sigma_N^2 = 0$!

These relations give some information on the correlation property we are interested in. In particular, white noise has infinite extent in the frequency domain, and therefore the autocorrelation function of the noise waveform has negligible extent in the time domain. This means that the correlation between noise voltages at two narrowly separated moments of time is very small.

The sampling theorem (Appendix E) also helps us to understand noise waveforms. It shows that a waveform with very wide bandwidth Δf in the frequency domain must be represented by a very large number of independent samples in the time domain, these being separated by intervals $1/2\Delta f$. Samples that are closer together than this interval will be correlated, but samples with greater than this separation can be taken as being independent. This indicates that a waveform that is subject to white noise can alternatively be considered as being subject to independent but identical* noise probability distributions at each distinguishable point in the time domain. This conclusion will be important in Chapter 15, where we take a sampled waveform that is subject to white noise, and assume that each temporal sample is subject to the same noise distribution with the same value of σ^2.

10.7   Noise in junction diodes†

There are two effects that allow charge carriers to conduct a current between the connections of a junction diode. The first is due to the (initially) majority carriers managing to surmount the potential barrier (e.g. electrons surmounting the barrier into the p-type region), an effect which increases current exponentially with the reduction of barrier height due to an applied (forward) voltage. The second is the leakage current that results from the (initially) minority carriers moving in the opposite direction. The well-known exponential law for the junction diode reflects these two opposing effects:

$$I = I_0[e^{eV/kT} - 1] \qquad (10.15)$$

Since the two currents described above both arise from the random motions of individual charge carriers, they are both subject to shot noise. In addition,

---

*In fact, it appears that noise whiteness leads to the independence property, but does not then *guarantee* that the individual temporal noise distributions will be identical: however, in practical situations, it will frequently be reasonable to make this assumption – as in the case of noise at the input of a radar receiver.

† This section may be omitted at first reading.

they have separate physical origins, and so they are uncorrelated noise sources (however, we shall see below that this statement requires qualification). We can therefore deduce that they give rise to a mean square noise current:

$$\overline{i_n{}^2} = 2eBI_0[e^{eV/kT} + 1]$$

$$= 4eBI_0 + 2eBI \qquad (10.16)$$

where we have used equation (10.15) to eliminate the explicit exponential variation. In fact, this is a quite revealing formula since it shows that there is a lot more (mean square) current noise when the diode is forward-biased than when it is reverse-biased (i.e. $\sim 2eI$ instead of $\sim 2eI_0$), while when the diode is switched off the noise is entirely due to the leakage current I_0.

Next, we consider the dynamic conductance of the device:

$$G = dI/dV = (I_0 e/kT)e^{eV/kT} = (I_0 e/kT)[1 + I/I_0] \qquad (10.17)$$

so that

$$kTG = e(I_0 + I) \qquad (10.18)$$

Using this equation to eliminate I_0 from equation (10.16), we find:

$$\overline{i_n{}^2} = 4B(kTG - eI) + 2eBI = 4kTBG - 2eBI \qquad (10.19)$$

This form of the equation is also very revealing. First, if $I = 0$, we get the normal Johnson noise formula corresponding to the *effective* resistance (or conductance) of the diode. However, in the forward direction, the shot noise is a lot less than might be expected – it is reduced from the value $4kTBG$ by the amount $2eBI$. Let us see what this amounts to in the case of strong forward conduction. In that case $kTG \approx eI$, so

$$\overline{i_n{}^2} \approx 2kTBG \qquad (10.20)$$

Note first that the coefficient of $kTBG$ has dropped by half compared with the situation when $I = 0$; and second, that G is now much lower than at $I = 0$. (The *dynamic* conductance is also much lower than the static d.c. conductance for strong forward conduction.) Thus the diode produces far less noise than any purely resistive circuit. From the arguments of Section 10.1, this can be taken to show that the diode would remove noise power from a resistor in the same circuit, and not replace it. Thus the resistor would be cooled: yet there is no doubt that the diode would be heated since it has a substantial I-V product. However, no laws of thermodynamics are broken by this effect, since the device is not in thermodynamic equilibrium with its surroundings: this is because its bias is being maintained, and power fed to it, from an exterior source of voltage.

Although the above calculation accurately predicts the observed low-frequency noise, it is based on premises which are not strictly correct, as the actual forward and reverse currents I_{fd} and I_{rd} flowing through the depletion region are very much larger than $I_f = I_0 \, e^{eV/kT}$ and $I_r = I_0$. It turns out that the large currents I_{fd} and I_{rd} are highly correlated and nearly cancel out, leaving just the smaller currents I_f and I_r – while the noise contributions also nearly cancel because of space-charge smoothing. Nevertheless, I_f and I_r are the actual macroscopic currents passing through the diode and must be subject to *at least* the noise levels given by the shot noise formula. However, what the shot noise formula does not predict is the rise in noise level at high frequencies; in addition, the full mechanism leading to the large forward and reverse currents I_{fd} and I_{rd} gives a sound explanation of the observed high level of saturation noise in the BJT. Thus the *detailed mechanism* by which diode noise arises is more to do with fluctuations in diffusion processes in the bulk semiconductor than to shot noise in the depletion region. For a full explanation of the rather complex situation see Buckingham (1983).

These facts are of considerable interest, but in practice it turns out that diode shot noise is relatively unimportant. This is because diodes are normally used only in parts of signal processing circuits where the signal is relatively strong. (There are good reasons for this, as we shall see, and these stem from the need for the action of the diode in a mixer or detector circuit to be as positive as possible, so that *incoming* noise is not mixed into the signal waveforms – see especially Chapter 18.) The main exception to this rule is in microwave mixers, which generally precede the main (i.f.) amplifier (low-noise microwave amplifiers are expensive to produce), and in that case the flicker noise generated in the diode is far more important than the shot noise.

10.8 Summary

This chapter has described the various types of noise that can arise in electronic systems. These include Johnson noise, shot noise, flicker noise, mains hum and pickup, r.f. and mains-borne interference, crosstalk and clutter. Johnson noise is thermal in origin and arises from resistive elements in circuits, while shot noise arises because of the discrete nature of charge which is responsible for carrying electric currents.

Like resistors, *ideal* junction diodes generate noise, and in fact most of the noise is calculable from shot noise formulae (though the *detailed mechanism* by which the noise arises is attributable to fluctuations in diffusion in the bulk semiconductor): the effect is normally less than the Johnson noise that might be expected on the basis of the diode's conductance. This is because the two shot noise processes tend to cancel in the forward direction, and as a result it turns out that the diode will cool other components it is electrically connected to! However, this summarizes the situation for ideal diodes, and in practice the flicker noise they produce is of greater importance than their shot noise output.

In addition to discussing the sources of noise in electronic systems, the chapter describes how overall noise levels may be calculated: for uncorrelated sources, the mean square noise voltages or currents should be added. However, when noise sources are correlated, the situation becomes more complicated and special formulae have to be used.

10.9 Bibliography

King (1966) is perhaps the most readable book on noise, but is now dated (see also Bell, 1960). Robinson (1974) is more up-to-date, but is rather hard work for the reader, while Buckingham (1983) gives crucial insight into noise processes in semiconductor diodes (see also Buckingham and Faulkner, 1974). Much work on noise is theoretically tedious, as it is statistical in nature, and books tend to quote results rather than deriving them: in this respect, note that Rice produced two seminal papers (1944, 1945), including studies of shot noise; see also van der Ziel and Becking (1958), van der Ziel (1960, 1962, 1963), and Robinson (1969a, b, c, d) for basic work on noise in semiconductor devices. Robinson (1969e) gives a penetrating insight into the distinction between shot noise and thermal noise processes. Finally, Baxandall (1968) gives a simpler introduction to noise and its properties, his article dating from a time when there was swift progress in understanding the noise characteristics of transistors.

10.10 Problems

1. Repeat the analysis of Section 10.2, assuming that a resistor generates a mean square *current* proportional to $h(R)$, and find the form that $h(R)$ must take. Relate this form to that of equation (10.3).

2. An a.c. meter employing a full-wave rectifier circuit is calibrated to display r.m.s. voltage when sinusoidal waveforms are input. Show that the meter will give an error $\sim 11\%$ when fed with square-waves and an error $\sim 13\%$ when fed with Gaussian noise. Determine the signs of the errors in the two cases.

3. A $10\,\text{k}\Omega$ resistor is connected across the input terminals of an oscilloscope of bandwidth 20 MHz. Assuming a value of kT of $4 \times 10^{-21}\,\text{J}$, determine the standard deviation of the Johnson noise voltage that would be expected to appear on the oscilloscope screen. Show that the noise will barely be visible on a scale of $1\,\text{mV/cm}$.

11

Noise in Amplifying Circuits

11.1 Introduction – the concept of noise figure

In many applications such as communications, amplifiers and other devices are connected together in such a way that maximum power transfer occurs. It is well known that they then have to be "matched" to each other, i.e. the output impedance of the one device is equal to the input impedance of the next. In addition, it is common to retain some standard impedance for all the devices in the chain, impedances of $50\,\Omega$, $75\,\Omega$ and $300\,\Omega$ being particularly common. Similarly, in developing the theory, it is useful to standardize on a particular impedance (which we shall call R_m), as this prevents ambiguity and at the same time simplifies the problems we shall try to solve. If at any stage a different impedance is required, e.g. to feed to a particular type of antenna or loudspeaker, there should be little difficulty in invoking a special impedance converter for this purpose.

In this context, it is useful to define the *noise figure* of an amplifier as the ratio of the signal-to-noise (power) ratios* between the input and the output:

$$F = \frac{(S/N)_i}{(S/N)_o} = \frac{S_i/S_o}{N_i/N_o} \tag{11.1}$$

The best possible noise figure is unity, since an ideal amplifier will amplify signal and noise by the same factor; on the other hand a non-ideal amplifier will introduce extra noise, making the signal-to-noise ratio (SNR) at the

---

* Elsewhere in this book it is most relevant to interpret SNR as a voltage ratio. However, in this chapter we take it to be a power ratio. In any case it is conventional to take noise figure as a ratio of power ratios, and thus it is natural to express it in dB.

141

output worse (i.e. lower) than at the input. Suppose we have an amplifier which has power gain G, and which generates internal noise N_o' which is added to the (amplified) input noise N_i. Then the output noise level is:

$$N_o = GN_i + N_o' = G(N_i + N_i') \qquad (11.2)$$

where we have re-expressed the internally generated noise as noise N_i' referred to the input terminals of the amplifier. In addition:

$$G = S_o/S_i \qquad (11.3)$$

so our expression for F can be rewritten in the form:

$$F = \frac{N_o/N_i}{G} = \frac{N_i + N_i'}{N_i} \qquad (11.4)$$

Hence

$$N_i' = N_i(F-1) \qquad (11.5)$$

Next, we are assuming that the amplifier is matched to its signal source, so the minimum possible input noise level is given by the noise N_m transmitted to the amplifier by a resistance of this source impedance, i.e. $N_i = N_m$ – a quantity which is equal to kTB, as we have already seen in Section 10.2. Thus

$$N_i' = N_m(F-1) \qquad (11.6)$$

(It should be remarked that N_m is the lowest possible value of N_i, so this leads to a value of F that is the highest possible, i.e. this is a worst-case definition: if we allowed the definition of F to apply when N_i included a high level of noise from a previous amplifier then F could easily be made to approach unity, and the definition would not be a useful one.)

11.2   Noise figure of a composite amplifier

It is pertinent to ask what the noise figure will be, for an amplifier constructed from two or more amplifiers connected in cascade. In the case of two amplifiers joined in cascade:

$$G = G_1 G_2 \qquad (11.7)$$

$$F = \frac{N_0/N_i}{G} = \frac{G(N_i + N_{i1}') + G_2 N_{i2}'}{N_i G} = \frac{(N_i + N_{i1}') + N_{i2}'/G_1}{N_i} \qquad (11.8)$$

where equation (11.5) shows that the internally generated noise powers referred to the inputs of the two amplifiers are:

$$N_{i1}' = N_m(F_1 - 1) \qquad (11.9)$$

and

$$N_{i2}' = N_m(F_2 - 1) \qquad (11.10)$$

Eliminating N_{i1}' and N_{i2}', and again setting $N_i = N_m$, we find:

$$F = \frac{N_m F_1 + N_m(F_2 - 1)/G_1}{N_m} = F_1 + \frac{F_2 - 1}{G_1} \qquad (11.11)$$

Note that if amplifier 2 is constructed from two further amplifiers in cascade, we have a similar formula for F_2, and the combined equation is obtained by substituting:

$$F_2 \rightarrow F_2 + (F_3 - 1)/G_2 \qquad (11.12)$$

so we get:

$$F = F_1 + (F_2 - 1)/G_1 + (F_3 - 1)/G_1 G_2 \qquad (11.13)$$

and in the general case the final (nth) term is:

$$(F_n - 1)G_n/G \qquad (11.14)$$

It is possibly more intuitive to express the general result in the form:

$$(F - 1) = (F_1 - 1) + (F_2 - 1)/G_1 + (F_3 - 1)/G_1 G_2 + \ldots \qquad (11.15)$$

since this expresses more emphatically that the added noise is the sum of the noise contributions introduced by the various stages, referred individually to the input of the overall amplifier.

Next we examine what happens when amplifiers are joined by lossy lines or attenuators. Suppose the attenuation factor is L, i.e. the signal level is divided by a factor L which is necessarily greater than unity. In fact, by definition

$$G = 1/L \tag{11.16}$$

The first part of equation (11.4) now gives:

$$F_{\text{atten}} = (N_o/N_i)L \tag{11.17}$$

and since attenuators are constructed from resistors which are intrinsically noisy, we may expect that F will be quite large. However, if the attenuator is ideal (in the sense that its resistances contribute nothing but Johnson noise) and is matched at its input and its output, then:

$$N_o = N_i = N_m \tag{11.18}$$

Hence:

$$F_{\text{atten}} = 1/G = L \tag{11.19}$$

The fact that F_{atten} is lower than we might originally have expected is due to our intuition which was based on a knowledge of the detailed implementation of an attenuator. The attenuator of Figure 11.1 is not a valid model of this device, a more realistic equivalent circuit being that shown in Figure 11.2. In the latter case, we have three parameters by which to adjust

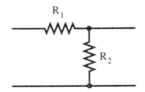

Fig. 11.1   Commonly used resistive attenuator.

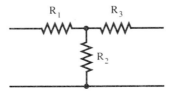

Fig. 11.2   Matched form of resistive attenuator. Adjustment of the three resistors will permit this attenuator circuit to be matched at both ends, while having any desired attenuation factor.

(a) the attenuation, (b) the input impedance, and (c) the output impedance. Attention to these three factors, followed by computation of the noise generated by all three resistors and attenuated by the remainder of the circuit *must* give the same result as above. It is left as an exercise for the reader to confirm this. The final result is determined by the fact that the attenuator is correctly matched at its input and output, so the appropriate *resultant* noise generators have resistance R_m.

11.3   Noise measure

We next consider the practical situation for a set of amplifiers of various gains and noise figures, where we need to determine the best order of connection in order to construct an optimal high-gain amplifier. First, take two amplifiers 1 and 2, and join them either way round. Then we have two overall noise figures:

$$F_{12} = F_1 + (F_2 - 1)/G_1 \qquad (11.20)$$

$$F_{21} = F_2 + (F_1 - 1)/G_2 \qquad (11.21)$$

If it is better to make amplifier 1 the first of the two amplifiers, then we will have $F_{12} < F_{21}$, so that:

$$F_1 + (F_2 - 1)/G_1 < F_2 + (F_1 - 1)/G_2 \qquad (11.22)$$

$$\therefore \quad (F_1 - 1)(1 - 1/G_2) < (F_2 - 1)(1 - 1/G_1) \qquad (11.23)$$

$$\therefore \quad \frac{F_1 - 1}{1 - 1/G_1} < \frac{F_2 - 1}{1 - 1/G_2} \qquad (11.24)$$

This makes it clear that amplifiers that have the lowest noise measure should be placed first in the sequence of amplifiers, where the *noise measure* of an amplifier is defined as its value of:

$$M = \frac{F - 1}{1 - 1/G} \qquad (11.25)$$

Note, however, that we have made an assumption in the above derivation – that the factor $(1 - 1/G)$ is necessarily positive. Clearly this is so for an amplifier, but it is not so for an attenuator. If *one* of the devices is an attenuator, then the deduced ordering is incorrect – i.e. any attenuator must if possible be *placed* after all the amplifiers in the chain. However, if *both* devices are attenuators, the situation theoretically reverts to the ordering deduced earlier. We should check this by deducing the noise measure of an attenuator:

$$M_{\text{atten}} = (L-1)/(1-L) = -1 \qquad (11.26)$$

Clearly, for *ideal* matched attenuators it actually makes no difference in which order they are placed – a fact that is fortunately in accord with intuition!

Next, we record an interesting property of noise measure. Suppose we have a long sequence of n identical amplifiers in cascade. Then the combined noise figure is:

$$F = F_1 + (F_1-1)/G_1 + (F_1-1)/G_1{}^2 + (F_1-1)/G_1{}^3 + \ldots + (F_1-1)/G_1{}^{n-1}$$

$$= 1 + \frac{(F_1-1)(1-1/G_1{}^n)}{1-1/G_1}$$

$$\approx 1 + \frac{F_1-1}{1-1/G_1} \qquad (11.27)$$

i.e.

$$F \approx 1 + M_1 \qquad (11.28)$$

This reflects the fact that the noise figure of a combined amplifier is minimized by placing first the amplifier with the lowest noise measure, and then the overall noise figure is greater than the ideal value of unity by an amount approximately equal to this lowest noise measure.

The above discussion shows that noise measure is a concept of some importance. It is also of interest that noise measure can be used not only for *ordering* of amplifiers, but also for *selection* of amplifiers from an arbitrary set (i.e. in a case where the desired gain is significantly less than the products of the gains of all the available amplifiers).

Finally, we note that the above analysis has been concerned with the *ideal* situation. In practical applications, cascades of amplifiers and attenuators cannot always be placed in anything like the ideal order. One obvious example of this is provided by observing that attenuating links often have to be placed between antennae and amplifiers and between one amplifier and another. Thus the placing of attenuating links is normally determined in advance, and the only freedom that remains is in the ordering of the amplifiers. However, careful analysis of any practical implementation on the lines indicated in the last two sub-sections should reveal which devices are the most important contributors to the noise figure: attention to these (be they cables, waveguides or amplifiers) should then permit relevant improvements to be made to the system as a whole.

11.4 Noise temperature

In Section 11.1 we studied the noise generated within an amplifier and showed how it affects the noise figure F. We now re-express equation (11.6) in the form:

$$N_i' = kTB(F-1) = kT_e B \qquad (11.29)$$

which suggests that the noise generated within the amplifier can be considered as due to a resistor at the input of the amplifier and at an *effective noise temperature* T_e. In that case

$$T_e = T(F-1) \qquad (11.30)$$

so that

$$F = 1 + T_e/T \qquad (11.31)$$

Here T is the temperature of the source feeding the amplifier, normally taken to be $T_0 = 290 \, \mathrm{K}$. This formula shows that an amplifier with a high noise figure has a high effective noise temperature. However, although the relation between F and T_e is linear, when F is close to unity, T_e may still be quite high, and indeed, T_e is a better measure of generated noise for high-quality amplifiers with low noise figures (see Table 11.1). In addition, the value of T_e is often related to some physical temperature in the amplifying system. For example, maser amplifiers operating at very low temperatures may have intrinsic noise temperatures of the same order (e.g. 4 K).

Unfortunately, the situation is seldom quite as simple as this, since waveguide or other transmission losses, which apparently add very little to F, can markedly increase the noise temperature. To show that this is so, consider the case of a lossy antenna feeding a maser amplifier. From equations (11.11), (11.16) and (11.19) we have:

$$F = L + (F_2 - 1)L = F_2 L \qquad (11.32)$$

Table 11.1   Noise figure and noise temperature values.
This table contains representative noise figure and noise
temperature values. It also forms a useful reference of
the decibel scale in terms of voltage and power ratios.

	Noise figure F		Noise temperature T_N
V ratio	P ratio	dB	K
1.000	1.000	0.000	0.00
1.002	1.005	0.022	1.45
1.005	1.010	0.043	2.90
1.007	1.015	0.065	4.35
1.010	1.02	0.086	5.8
1.015	1.03	0.128	8.7
1.020	1.04	0.170	11.6
1.025	1.05	0.212	14.5
1.034	1.07	0.294	20.3
1.049	1.1	0.414	29
1.095	1.2	0.792	58
1.140	1.3	1.139	87
1.183	1.4	1.461	116
1.225	1.5	1.761	145
1.265	1.6	2.041	174
1.342	1.8	2.553	232
1.414	2.0	3.010	290
1.581	2.5	3.979	435
1.732	3	4.771	580
2.000	4	6.021	870
2.236	5	6.990	1160
2.646	7	8.451	1740
3.162	10	10.000	2610
10.000	100	20.000	28710

Substituting for F and F_2, we find:

$$T_e = T_0[(1 + T_{e2}/T_0)L - 1] = (L - 1)T_0 + LT_{e2} \qquad (11.33)$$

Thus the noise temperature is higher than T_{e2} for *two* reasons: the first is
that the loss factor of the attenuator cuts down the amount of signal reaching
the maser, thereby making the second term greater than T_{e2}; the second is
that the attenuator introduces a small proportion of the much higher

temperature T_0, and if $L \approx 1.05$, we find:

$$T_e \approx 0.05 \times 290 + 1.05 \times 4 = 18.7\,\text{K}$$

which is very much greater than the intrinsic value of 4 K.

Physically, it is intriguing how the lossy antenna can add to the noise temperature. In fact the mechanism is quite simple. Imagine a speck of dirt on the reflector. Clearly, it prevents some of the signal from arriving at the receiver by obscuring that part of the reflector. However, the speck of dirt is at temperature T_0: it therefore transmits black-body radiation of that temperature into the receiver, and thereby gives rise to the $(L-1)T_0$ term in equation (11.33)*. This example shows the value of the noise temperature concept in clarifying some of the issues relating to receiver sensitivity.

We finish this section by rewriting equation (11.15) in the form:

$$T_e = T_{e1} + T_{e2}/G_1 + T_{e3}/G_1G_2 + \ldots \qquad (11.34)$$

The somewhat greater simplicity of this equation underlines the comments made immediately following equation (11.15).

11.5 Noise equivalent bandwidth

The low-pass RC filter has magnitude voltage attenuation that varies with frequency in the following way:

$$\frac{v_o}{v_i} = \frac{1/\omega C}{[R^2 + (1/\omega C)^2]^{1/2}} \qquad (11.35)$$

so the power attenuation factor is:

$$P_o/P_i = 1/[1 + (\omega CR)^2] = 1/[1 + (f/f_0)^2] \qquad (11.36)$$

where

$$f_0 = 1/2\pi RC \qquad (11.37)$$

At $f = f_0$, $P_o/P_i = 1/2$, so attenuation $\approx 3\,\text{dB}$

At $f = f_0/2$, $P_o/P_i = 4/5$, so attenuation $\approx 1\,\text{dB}$

At $f = 2f_0$, $P_o/P_i = 1/5$, so attenuation $\approx 7\,\text{dB} = 6\,\text{dB} + 1\,\text{dB}$

---

* Note that a small hole in the antenna would have an almost identical effect!

Thus the straight-line approximations to the performance of the filter underestimate the signal attenuation by 3 dB at f_0, but only by 1 dB at $f_0/2$ and $2f_0$, these latter two frequencies being equally spaced from f_0 on the logarithmic frequency scale of Figure 11.3.

We now turn to the noise performance of the filter. The *noise equivalent bandwidth* B_N is defined as the bandwidth of an ideal filter of rectangular frequency response which will transmit the same noise power as the actual noise filter under white noise conditions. Thus:

$$B_N = \int_0^{B_N} df = \int_0^\infty \frac{1}{1+(f/f_0)^2} df = f_0 \int_0^\infty \frac{1}{1+u^2} du \qquad (11.38)$$

where

$$u = f/f_0 \qquad (11.39)$$

$$\therefore \quad B_N = f_0[\arctan u]_0^\infty = (\pi/2)f_0 \qquad (11.40)$$

Substituting for f_0, we find:

$$B_N = (\pi/2)/2\pi RC = 1/4RC \qquad (11.41)$$

Thus the noise bandwidth B_N is 1.57 times as large as the signal bandwidth B – or in other words, the effective noise cut-off frequency is 1.57 times the signal cut-off frequency f_0 (see Figure 11.3).

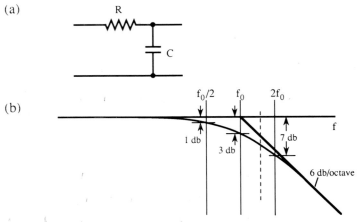

Fig. 11.3   *R-C* low-pass filter and its frequency response. (a) An *R-C* low-pass filter, and (b) its frequency response. Above the cut-off frequency f_0, the attenuation falls off (in the limit) at 6 dB per octave (20 dB/decade). The dotted line shows the frequency corresponding to the noise bandwidth of the circuit (see text).

11.6   Noise in electronic circuits

Electronic devices such as valves and transistors generate various forms of noise including in particular Johnson noise and shot noise. As we have seen, Johnson noise arises as thermal noise in resistances, while shot noise arises as fluctuations in current due to the electronic charge. Clearly, both are fundamental and unavoidable – being present even in ideal devices: hence the best that can be done is to minimize these effects, by careful design of both the devices and the circuits that include them. Flicker noise is another form of noise that is probably less fundamental, and which appears to arise from surface effects on solid materials. However, most existing solid state devices introduce significant flicker noise, so again attempts must be made to design circuits in such a way as to minimize its effects. Unfortunately, flicker noise is "pink" ($1/f$) noise, and therefore it has relatively large effects at low frequencies (particularly below 1 kHz). Thus in many circuits it is of much greater importance than the white noise arising from Johnson and shot noise.

We have already seen that, in a cascade of amplifiers, the noise characteristics of the first amplifier in the chain are of greatest importance, and it is normal to characterize the noise performance of an amplifier by the noise power referred to the *input* of the amplifier. For a single amplifying device, such as a transistor, we can therefore imagine that all the noise produced inside the device is normalized to a number of voltage and current noise generators at its input connection.

11.6.1   Modelling noise for the BJT

We now examine these sources of noise in detail in the case of the BJT. First, the device will be fed from a voltage source of internal resistance R_s, and this will lead to a mean square noise voltage:

$$\overline{v_{n,s}{}^2} = 4kTR_sB \qquad (11.42)$$

Next, consider the current passing through the device. If this is I, then it will be subject to shot noise, which is given by equation (10.5). However, we must normalize this to the input of the device by *dividing* the noise current by the mutual conductance y_f; this yields an equivalent mean square input noise voltage:

$$\overline{v_{n,I}{}^2} = 2eIB/y_f{}^2 \qquad (11.43)$$

For a BJT, it is useful to recall that

$$y_{fe} \approx eI_c/kT \tag{11.44}$$

so that

$$\overline{v_{n,I}^2} = 2kTB/y_{fe} \tag{11.45}$$

There is also a small noise component from the series base resistance r_b of the transistor; this leads to the modified total mean square noise voltage:

$$\overline{v_n^2} = 4kTB(r_b + 1/2y_{fe}) \tag{11.46}$$

Finally, there is a shot noise component due to the base current I_b where the latter may be estimated by dividing the collector current I_c by h_{fe}. Thus we get:

$$\overline{i_{n,b}^2} = 2eI_cB/h_{fe} = 2kTBy_{fe}/h_{fe} \tag{11.47}$$

where we have again applied equation (11.44). (Note also that $y_{fe}/h_{fe} = 1/h_{ie}$, though this simplification will not be required in what follows.)

It is now useful to consider all these noise components as coming from a single noise voltage generator. To achieve this we have to consider the base shot noise current source as emerging from a noise voltage source which is in series both with the input voltage of internal impedance R_s and with r_b. Hence we have to multiply its value by $(R_s + r_b)^2$ to obtain the appropriate mean square voltage:

$$
\begin{aligned}
\overline{v_n^2} &= 4kTR_sB + 4kTB(r_b + 1/2y_{fe}) + 2kTB(R_s + r_b)^2 y_{fe}/h_{fe} \\
&= 4kTB[R_s + (r_b + 1/2y_{fe}) + (R_s + r_b)^2 y_{fe}/2h_{fe}]
\end{aligned} \tag{11.48}
$$

11.6.2   Minimizing the effects of noise

We now need to minimize the effects of noise. In fact, straight minimization of noise is in principle invalid since the means of achieving this might affect the gain of the amplifier, and it is the SNR that must eventually be optimized. This means that we should consider the noise figure F for the stage. As normally defined, F is the factor by which the SNR is degraded by the amplifier, and this factor is equal to the effective noise input power divided by that of the equivalent ideal amplifier (see equation 11.4):

$$F = \text{effective input noise power/ideal input noise power}$$

$$= \frac{4kTB[R_s + (r_b + 1/2y_{fe}) + (R_s + r_b)^2 y_{fe}/2h_{fe}]}{4kTBR_s}$$

$$= 1 + \frac{r_b + 1/2y_{fe}}{R_s} + \frac{(R_s + r_b)^2 y_{fe}}{2h_{fe}R_s} \tag{11.49}$$

There are two ways we can optimize this expression in order to minimize noise. The first is to note that y_{fe} is proportional to I_c, so altering the working conditions of the transistor will permit noise to be minimized. To achieve this we find the derivative:

$$dF/dy_{fe} = -1/2R_s y_{fe}^2 + (R_s + r_b)^2/2h_{fe}R_s \tag{11.50}$$

Setting this equal to zero now gives:

$$y_{fe} = h_{fe}^{1/2}/(R_s + r_b) \tag{11.51}$$

or, alternatively

$$I_c = \frac{kTy_{fe}}{e} = \frac{(kT/e)h_{fe}^{1/2}}{R_s + r_b} \tag{11.52}$$

The minimum value of F achievable in this way is:

$$F_{min} = 1 + r_b/R_s + (R_s + r_b)/R_s h_{fe}^{1/2} \tag{11.53}$$

The second way of minimizing the noise is to consider the overall noise figure of the stage and to adjust the value of R_s for an optimum. To carry this out it is convenient to define the following quantities:

$$R_v = r_b + 1/2y_{fe} \tag{11.54}$$

$$R_i = 2h_{fe}/y_{fe} \tag{11.55}$$

and then to re-express the noise figure in terms of them. Here, R_v is the equivalent resistance noise generator for the noise *voltage* sources, and R_i is the equivalent resistance noise generator for the noise *current* sources. R_i is deduced using the relation

$$\overline{i_{n,i}^2} = \overline{v_{n,i}^2}/R_i^2 = 4kTR_iB/R_i^2 = 4kTB/R_i \tag{11.56}$$

which in this case has to equal $\overline{i_{n,b}^2}$, as given originally by equation (11.47). At this point we simplify the mathematics by assuming that $r_b \ll R_s$, which is normally a good approximation, since r_b is likely to be lower than $200\,\Omega$ (see below) and R_s is typically greater than $1\,\mathrm{k}\Omega$. We now obtain equation (11.49) in the simplified form:

$$F = 1 + R_v/R_s + R_s/R_i \tag{11.57}$$

$$\therefore \quad dF/dR_s = -R_v/R_s^2 + 1/R_i \tag{11.58}$$

Setting this expression to zero in order to optimize the SNR gives:

$$R_s = R_{so} = (R_v R_i)^{1/2} = [(1 + 2y_{fe}r_b)h_{fe}/y_{fe}^2]^{1/2} \tag{11.59}$$

Substituting for R_s in the formula for the noise figure now gives:

$$F_{min} = 1 + 2(R_v/R_i)^{1/2} = 1 + [(1 + 2y_{fe}r_b)/h_{fe}]^{1/2} \tag{11.60}$$

Unfortunately, the designer often has little control of R_s. Thus it is of interest to determine how much the noise deteriorates if R_s is *not* given by equation (11.59). In fact the situation is helped considerably by the fact that the variation around the calculus minimum of F is necessarily slowly varying, and indeed for small values of R_v/R_i the curves giving the variation in F are quite shallow (see Figure 11.4): thus it is not at all necessary to insist on being close to the optimum value of R_s. To understand this in more detail, we express F in terms of $\eta = R_s/R_{so}$:

$$F = 1 + (R_v/R_i)^{1/2}(\eta + 1/\eta) \tag{11.61}$$

which remains unchanged as η is replaced by $1/\eta$, indicating that the variations will be symmetrical on a logarithmic plot of F against η. In particular, if η is increased or decreased by a factor 10 from the optimal value $\eta_o = 1$, we have:

$$F = 1 + (R_v/R_i)^{1/2}(10 + 1/10) \approx 1 + 5(F_{min} - 1) \tag{11.62}$$

Taking for example the case $R_v \approx 300\,\Omega$, $R_i \approx 300\,\mathrm{k}\Omega$, for which $R_{so} \approx 10\,\mathrm{k}\Omega$, $F_{min} \approx 1.07$, and a factor 10 variation in R_s gives $F \approx 1.35$. (Note that these noise figure values are power ratios, and correspond to $0.3\,\mathrm{dB}$ and $1.3\,\mathrm{dB}$ respectively.)

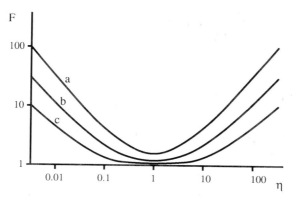

Fig. 11.4 Variation in noise figure around the optimum position. This figure shows how the noise figure F varies around the optimum position, η being the quantity R_s/R_{so}. Curves (a), (b), and (c) show the respective cases $R_i/R_v = 10$, 100 and 1000.

Finally, note that F_{min} has a limiting value of $1 + 1/\sqrt{h_{fe}}$ as $r_b \to 0$. This shows the importance of finding devices with low values of r_b, and high values of h_{fe}, for low-noise work. Although typical transistors have values of r_b in the range 200–1000 Ω, transistors designed for low-noise applications have values of a few *tens* of ohms, and some have values as low as 4 Ω.

11.6.3   Noise in FETs

The above discussion started quite generally, and then became specific to BJTs. It is of great relevance to understand how the picture presented above becomes modified with FETs. First, equation (11.42) remains valid but equation (11.43) no longer applies, since Johnson noise dominates shot noise in the FET channel. (The end result of this is that equation (11.45) applies, but with a numerical coefficient of $\frac{8}{3}$ instead of 2.) In addition, the Johnson noise from r_b no longer exists. Finally, there is very little gate noise current, so the shot noise from this source can often be ignored. This means that as a *first approximation* R_i in equation (11.57) can be taken as infinite and the term containing it ignored. Hence R_{so} is very high and it is difficult to satisfy the condition for optimum F. On the other hand, the curves for variation of F with R_s are again very flat, since typically $R_v \approx 100\,\text{k}\Omega$ and $R_i \approx 40\,\text{M}\Omega$, giving $(R_v/R_i)^{1/2} \approx 0.05$ and $R_{so} = (R_v R_i)^{1/2} \approx 2\,\text{M}\Omega$. Broadly, these figures indicate that the BJT will give better noise performance for low values of R_s and the FET for high values of R_s, the break-point perhaps being around 100 kΩ. However, this simple picture is modified considerably if flicker noise and frequency-dependent gain factors are taken into account. We now proceed to consider flicker noise.

11.7   Flicker noise contributions

The discussion in the previous section was concerned primarily with Johnson and shot noise, both of which are fundamentally white. To these must be added flicker noise, which has a $1/f$ spectrum and is particularly strong below 1 kHz. However, it is not clear *a priori* where flicker noise components should be added in the formulae derived earlier. In fact, for the BJT, it turns out that flicker noise contributes mainly to the base current shot noise, thereby augmenting the $\overline{i_{n,b}{}^2}$ current generator:

$$\overline{i_{n,b}{}^2} = 2eI_cBh_{fe}{}^{-1}(1+f_c/f) = 2kTBy_{fe}h_{fe}{}^{-1}(1+f_c/f) \qquad (11.63)$$

where f_c is the flicker noise "corner" frequency (Figure 11.5). However, it must be remembered that modelling flicker noise and adding it in this way is only an approximation, and makes assumptions about the lack of correlation of the flicker noise and white noise sources. Using this equation, we can correct equation (11.52), giving the optimum collector current as:

$$I_c = \frac{kTy_{fe}}{e} = \frac{(kT/e)[h_{fe}/(1+f_c/f)]^{1/2}}{R_s+r_b} \qquad (11.64)$$

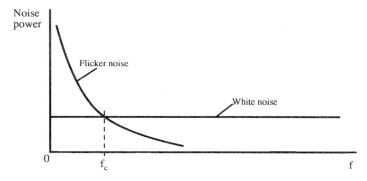

Fig. 11.5   Frequency spectra of white noise and flicker noise. This figure shows the frequency spectra of white noise and flicker noise, and effectively defines the flicker noise "corner" frequency f_c.

indicating that the increased flicker noise at low frequencies is partly offset by reducing the collector current. With a similar derivation, the equation for R_i now becomes:

$$R_i = \frac{2h_{fe}}{y_{fe}(1 + f_c/f)} \tag{11.65}$$

which is clearly reduced by the presence of flicker noise – with the result that R_{so} is also reduced, being divided by the factor $(1 + f_c/f)^{1/2}$.

For the FET, flicker noise arises predominantly as an increase by the factor $(1 + f_c/f)$ in the channel noise power at low frequencies. In this case the optimum source resistance R_{so} is *increased*, this time being *multiplied* by the factor $(1 + f_c/f)^{1/2}$.

An important further difference between BJTs and FETs is that the value of f_c is typically 1 kHz for the former and 20 kHz for the latter. This implies that FETs will perform less well as low-noise devices at low frequencies. It should also be pointed out that insulated gate FETs generate significantly more flicker noise than junction-gate FETs, and this renders them far less suitable for low-noise amplifiers.

Next, it is interesting that dual-gate FETs have been constructed for various purposes – including use in mixers and high-frequency amplifiers. In the latter case the additional gate is grounded and has the effect of reducing the capacitance between the first gate and the drain, thereby suppressing the Miller effect (see Chapter 4). In the earlier thermionic valve technology, the additional grids in the tetrode and pentode led to a proportion of the electrons falling on these extra grids and producing significant additional shot noise which was called *partition noise*. It turns out that there is no partition noise with dual-gate FETs, presumably because negligible numbers of carriers are lost to these connections, and there is therefore no current that can give rise to additional shot noise.

Finally, it is worth pointing out that device characteristics supplied by manufacturers frequently quote *spot noise* figures: these are mean square voltages per unit bandwidth at various "spot" frequencies, and the quoted units are r.m.s. volts per root-hertz. Flicker noise characteristics are generally not quoted directly, and have to be deduced from the spot noise figures quoted by manufacturers.

11.8   The complexity of noise calculations

Complete noise calculations can be very intricate and tedious to carry out – especially where (a) high-frequency characteristics of the devices (including the fall-off in gain at high frequencies) have to be taken into account, (b) amplifiers are tuned, (c) impedance matching is to be achieved, or (d) feedback over one or more stages has to be undertaken.

We can get some flavour of the complexity of these problems by considering what happens when one amplifier has to be joined to another and matched for optimum power transfer. Suppose that the second amplifier demands an ideal source resistance R_{so} so that its noise performance can be optimized, and R_{so} is not equal to the output impedance of the first amplifier (or that of the transmission line joining them): in such cases there can be a serious problem, but it can be tackled in a variety of ways. First, note the possibility that the second amplifier could use an FET instead of a BJT amplifying device at the input (this would have the effect of *raising* R_{so}), in order roughly to perform the noise matching. Note that this matching need not be carried out exactly, since the noise figure curves (F v. η) tend to be quite shallow. Second, if these considerations have not already solved the problem, it may be possible to employ transformer coupling to perform impedance matching, thereby raising *or* lowering R_{so} (but note that this approach is not quite so practicable if d.c. signals or pulses have to be transmitted). Third, parallelizing input transistors in the second amplifier can also help with this task (this would have the effect of *lowering* R_{so}). Finally, feedback methods can be used to adjust the input or output impedance to help with the impedance matching task.

The problems outlined at the head of this section, and the potential solutions listed in the last paragraph, are considered in detail in more specialized texts, such as Robinson (1974), and are beyond the scope of the present volume. Suffice it to say that noise optimization is not undertaken lightly. However, it is frequently an adequate policy to examine the characteristics of any given design and to isolate the *major* factor influencing noise; then to look out for a specialized commercial circuit or device that can be used to overcome this local problem. For example, an antenna amplifier may be used to boost signals at an early stage, so that the noise problem becomes far less serious, and then good circuit practice may well be sufficient to eliminate further problems.

11.9 Low-noise amplifiers in radio astronomy

In radio astronomy and satellite communications, antenna gain is a vital factor, but eventually the signals must be handled electronically, and it is necessary to employ low-noise amplifiers in the early stages. For this purpose, masers and parametric amplifiers (see Appendix C.1) used to be essential, since they had the lowest available noise figures and temperatures. However, the situation is not static. First, masers have the lowest noise of all amplifiers, with noise temperatures as low as 2 K, but they are expensive and tedious to use, and have to be cooled with liquid helium. Thus satellite receiver ground-stations have in many cases replaced them with parametric amplifiers. For best performance, the latter also have to be cooled, but noise temperatures ~ 50 K can be achieved using liquid nitrogen coolant. Second, modern GaAs FETs have remarkably good noise performance, with noise figures ~ 1.5 dB (noise temperature ~ 120 K) or significantly lower if they are cooled with liquid nitrogen (77 K). In addition, custom assembly and tuning of high electron mobility transistors (HEMTs) – which are actually GaAs FETs – can lead to noise temperatures lower than 10 K when cooled: these devices have recently been found invaluable in radio astronomy.

11.10 Summary

This chapter has discussed noise measures of various sorts, and means by which noise may be minimized in amplifier circuits. First, *noise figure* was defined as the ratio of the SNR at the input to that at the output: this quantity is always greater than unity, since amplifiers will add extra noise to the input signal. Then it was shown how the noise figure of a composite amplifier could be calculated. In practice, the *noise measure* of an amplifier is often a more relevant quantity, since it can be used to determine the priority order for selection between a number of amplifiers, and the optimum order for placing them in cascade.

The chapter also defines the noise equivalent bandwidth of an R-C filter – a quantity which turns out to be greater than the normal (3 dB) signal bandwidth.

Section 11.6 models the noise produced by an amplifier and fed to it by a source of given output impedance. The result of calculations made on this basis is that there is a point at which the noise figure is a minimum. Although the optimization condition is not specific to BJTs or other amplifying devices, when numerical values are inserted, the conditions vary considerably between (for example) BJTs and FETs. As a result, BJTs are expected to give better noise performance for low values of source resistance, and FETs are expected to be more suitable for high values of source resistance, the break-point being around $100\,k\Omega$. However, this basic model ignores flicker ($1/f$) noise and frequency-dependent gain factors. When frequency is brought into the calculations, it is found that FETs perform less well at low frequencies, since they have corner frequencies for flicker noise $\sim 20\,kHz$ compared with $\sim 1\,kHz$ for BJTs.

Unfortunately, noise calculations can be exceedingly complex (a) at high frequencies, (b) where tuned amplifiers are employed, (c) where impedance matching is required or (d) where feedback is involved. Such cases are beyond the scope of the present text, and the reader is referred to other sources (e.g. Robinson, 1974) for further information.

11.11   Bibliography

King (1966) provides a useful reference on the work of this chapter. Watson (1989) includes an up-to-date but fairly theoretical discussion on noise in amplifiers, while Horowitz and Hill (1989) cover the topic very thoroughly at a practical and design level. Robinson (1974) goes into considerably more depth, especially on the effects of feedback, use of transformer coupling to match impedances, and other topics referred to in Section 11.8. An excellent review of the subject appears in Netzer (1981). Jolly (1967) covers the concept of noise temperature, and enters a discussion of the principles, merits and characteristics of masers and parametric amplifiers: see also Buckingham (1983) and, at a more elementary level, Connor (1973). For practical aspects of satellite communications and radio astronomy, see Gould and Lum (1975), Edelson (1977), Bhargava et al. (1981), and Kitchin (1984). The reader is also referred to the early papers on noise cited at the end of Chapter 10.

11.12   Problems

1. Define noise figure.
   In a system composed of an amplifier and an attenuator, the attenuator is to be adjusted so that the overall power gain of the system is reduced to a predetermined value G. Obtain a formula for the noise figure of the system. Deduce from your formula how to select the best amplifier for an application where a power gain of G is required. Carry out this

procedure for the following set of amplifiers, if G is to be 4:

Amplifier	Power gain	Noise figure
a	3	1.1
b	4	2.0
c	5	1.7
d	10	1.8
e	15	1.6
f	25	1.9

Discuss briefly whether it is worth using more amplifiers than are strictly necessary, and then cutting down the excess gain by means of attenuators.

2. Explain why it is wrong to match an amplifier of low R_{so} (see equation 11.59) to the voltage source feeding it by inserting a series resistor of value $R_{so} - R_s$ (or any other value) between them.

3. Show that if an amplifier has n transistors in parallel in its input stage, its value of R_{so} (equation 11.59) is reduced by a factor n.

4. A submarine cable contains n repeaters and $n-1$ sections of attenuating cable. Assuming that each repeater contains an amplifier of noise figure F whose gain G compensates for the loss in one section of cable, obtain a formula giving the overall noise figure of the link. Consider also the reasons for not placing an amplifier of gain G^n at the beginning or end of the link. If a means were found for producing an amplifying cable whose gain were uniformly distributed along its length, determine whether this would constitute an ideal solution to the problem.

Part 3
Signal Recovery

This part of the book considers signal recovery – the science of extracting signals that are embedded in noise. The topic is introduced in Chapter 12, where it is seen that a.c. detection can be of value if a suitable range of frequency-domain filters is available. Chapter 13 discusses how ultra-narrow band filtering can be achieved by phase-sensitive detection techniques, and Chapter 14 extends the time-slicing concept to the use of boxcar detectors, but under the broader heading signal *averaging*. Chapter 15 develops the boxcar detection concept to include matched filtering and correlation. Then Chapters 16 and 17 outline two major application areas for these pulsed signal recovery methods – radar and magnetic spin-echo systems. Although these topics might seem rather specialized, radar has been important historically (and continues to be important) for signal recovery, while magnetic resonance exemplifies the application of signal recovery in quite wide areas of experimental physics and chemistry – not to mention the recently developed magnetic resonance imaging body scanning technology. Chapter 18 provides background on radio technology, and discusses in detail certain important issues on detection and mixing – in particular how noise reacts to these processes. Chapter 19 deals with certain advanced topics on signal recovery that have been skated over in earlier chapters, while Chapter 20 dwells on elimination of noise in the digital images that can arise in (for example) radar and medicine. Chapter 21 attempts to summarize what has been achieved in Part 3, not least by placing it in relation to the broader field of signal processing.

This part of the book necessarily embodies certain infelicities in the ordering of material. Readers will vary radically in their background knowledge – do they know about radio, noise, microwaves, digital electronics, computers or physics? On the other hand, the topic of signal recovery cuts right across the subject of electronics, making it difficult to explain its concepts at the same time as explaining its applications (for example, a full appreciation of the superheterodyne receiver cannot be obtained before the effects of mixing and noise have been considered in detail, though the basic idea is quite simple). This means that readers may well have to skip back and forth over the chapters of Part 3, and over the appendices, according to their immediate needs. However, every attempt has been made to facilitate this task by suitable cross-referencing and indexing.

Finally, an important detail is that we shall be most interested in voltage output variations in signals, and shall commonly use S and N to denote signal voltage and noise *r.m.s.* voltage respectively. In addition, SNR will normally be expressed as a voltage ratio. This contrasts with the treatment in Part 2, where these parameters arose as powers and power ratios.

12

Introduction to Signal Recovery

12.1 Introduction

The subject known as signal recovery is concerned with recovering signals from a background of noise. Sometimes signals are subject to a light sprinkling of noise and then signal recovery is in principle a relatively simple process. On other occasions, signals are bathed in so much noise that they are completely submerged – i.e. the signal-to-noise ratio (SNR) is less than unity – and it is impossible to detect their presence, let alone measure them and obtain useful information from them: in that case signal recovery can be a very complex task, and it may even be impossible to achieve. Between these two extremes there is a whole range of situations, and often considerable skill on the part of the electronics system designer is required if the process of signal recovery is to be successful. We shall see that there is a variety of techniques which can be considered while constructing such a system. In many cases one of the basic techniques will work unaided. More often a combination of techniques is required if the best results are to be achieved. We shall see that much depends on the nature of the signals, on the nature of the noise to which they are subjected, and whether for instance the signals can be coded in such a way as to make them easier to discriminate from noise.

12.2 Frequency domain filtering

To start with we assume that the signals have been amplified to a convenient level, so that they would no longer be too small to be visible if the noise were removed. Next, we look for some characteristic of the signals by which they might be distinguished from noise. In a large number of cases the signals appear at a certain frequency, or with a particular somewhat restricted frequency range. Hence it becomes useful to employ various types of frequency domain filter to help eliminate the noise. For signals that are restricted to low frequencies, a low-pass filter may be all that is required. In other situations high-pass filters may be useful. Perhaps more commonly, if signals appear around a fixed frequency, a band-pass filter will be required. There are variations on this, such as a narrow-band filter if the range of signal frequencies is especially narrow. It is somewhat difficult to be more specific without considering particular applications in detail. For example, if a signal is an amplitude-modulated wave centred around 1 kHz, a filter of quite narrow bandwidth – perhaps as low as 30 Hz (corresponding to a Q of about 30) – may be useful: this can be constructed using a twin-tee type of feedback circuit. On the other hand, if the centre frequency is 1 MHz, tuned circuits are likely to provide a better solution: this remains so in one form or another right up to microwave frequencies, though tuned circuits in that case are more likely to be resonant cavities than "lumped" component circuits (note that the Q-factor of a tuned circuit operating at 1 MHz might be ~ 200, whilst the Q for a resonant cavity operating at 10^{10} Hz might be 5000 or higher).

Another situation that frequently arises is when the noise is structured in some way. For example, there may be considerable interference at 50 Hz or 100 Hz, from power supply lines and so on. In that case a band-stop or band-reject filter is required (a narrow-band-reject filter of this type is often called a "notch" filter). So far we have considered unstructured noise and interference at a specific frequency. There are also other forms of noise to be considered. In particular, noise may not be "flat" over a wide frequency range, i.e. "white", but may be coloured: for example, "pink" noise is noise that is accentuated at the low-frequency end of the scale. However, some types of noise are accentuated at fairly high frequencies – as in the case of interference from car ignition systems or from lightning – though these latter types of noise do not have well-defined frequency spectra.

In fact, flicker noise is one of the most problematic forms of noise, and has been found to have a $1/f$ (pink noise) spectrum over a very wide range. Various electronic devices seem to produce this form of noise, often as a result of the unusual surface properties of some of the solid materials used in such devices. We do not need to examine the causes of this type of noise here. Suffice it to say that $1/f$ noise has been detected all the way from d.c. to microwave frequencies, and it is one of the primary problems with which we have to deal. The fact that its amplitude is reduced as frequency rises gives us a very positive incentive to modulate signals that would otherwise be steady, and to detect them at some convenient high frequency rather than at d.c. This is a form of coding that is commonly applied in order (as remarked

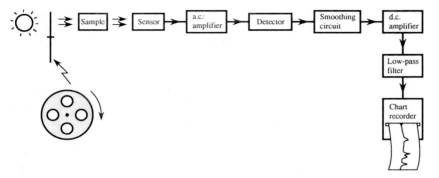

Fig. 12.1   a.c. optical detection system. This figure shows the apparatus required for an a.c. optical detection system. The chart recorder is intended to draw out some optical characteristic of the sample, such as the variation in response with frequency. Similar equipment would be required if the variation in amplitude with applied magnetic or electric field or temperature were to be plotted. The important feature of the apparatus is that a filter is applied at some frequency where flicker noise is relatively low.

above) to help discriminate signals from noise, thereby aiding detection and measurement. Thus we replace the obvious (d.c.) form of signal detection system with the a.c. system shown in Figure 12.1. Of course, this leaves us with an a.c. signal, and this then has to be rectified (or "detected") in order to reproduce the required d.c. signal – as indicated in Figure 12.1.

12.3   Ultra-narrow band systems

The next problem is that the noise level might not be reduced far enough by the above process. In particular, if the a.c. signal at the output of the narrow-band filter is still bathed in significant noise (i.e. the SNR is still of the order of unity, or lower), then the SNR after rectification will be very poor or even negligible, and in any case the output signal will not be proportional to the input signal*. In that case what is needed is a *very* narrow-band filter. However, there are problems in designing stable narrow-band filters, as indeed there also are in maintaining a signal at a

---

* Note that a simple detector diode does not operate linearly when the SNR is small – for a fuller discussion see Chapter 18.

constant frequency. In fact, the problem is more profound than this: if information is to be obtained from the signal at a reasonable rate, then the signal will be modulated and will have sidebands*. Narrowing the bandwidth of the output filter will tend to eliminate these sidebands and hence will remove the information we are trying to obtain. This problem is ultimately insoluble. However, we can at least attempt to design a system which will permit the output signal to be proportional to the input signal level and not to be distorted or randomly attenuated by noise. To achieve this a synchronous detection system is called for. This approach will be described in the next chapter. Meanwhile, we note that when conventional filters are to be used, it is usual† to employ a low-pass filter (often a simple time-constant) in or after the final d.c. amplifier: this will be easier to adjust than an a.c. filter – mainly because with the latter both the bandwidth and the centre frequency have to be adjusted. Thus the a.c. narrow-band filter would perhaps have a bandwidth of 30 Hz centred on 1 kHz, whereas the low-pass filter might have a bandwidth ~ 5 Hz – a value that can be brought down to 0.1 Hz or less if a synchronous detection system is used.

12.4   Summary

This chapter has examined the problem of how to recover signals that are buried in noise. A generally useful approach to signal recovery is to consider the signals and noise in the frequency domain, since it is often possible to apply frequency domain filters to this task, with great effectiveness. Another useful approach is to code the signals in a special way to make them easier to discriminate from noise. Both approaches are illustrated by the technique of modulating the signal at a frequency where flicker noise is significantly reduced, and then applying a narrow-band filter and rectifying the a.c. waveform.

Unfortunately, it can happen that the need to increase SNR cuts down the bandwidth so much that signal variations which carry important information are suppressed – a problem that can only be overcome by

---

* Sidebands are frequency components at nearby frequencies that can be used to carry additional information – see Appendix D.
† However, this does not automatically give an optimal solution: it can be shown that when an a.c. signal is to be detected and its bandwidth is restricted by filters to B before detection, and b after detection, then the final *noise* bandwidth is proportional to $\sqrt{2Bb}$ (see Chapter 18). This means that for an optimal solution, both an a.c. *and* a d.c. filter are required.

extending the time to make a particular set of observations. An additional problem, which will be tackled in the following chapter, is that it may be very difficult to design a suitable ultra-narrow band filter that has adequate stability.

12.5 Bibliography

There are a number of sources that can help with the study of frequency domain filtering, and more generally with the subject of signal recovery. First, the book by Robinson (1974) on noise and fluctuations forms a basic text; second the title by Wilmshurst (1985) on signal recovery is especially relevant, and covers many interesting topics on signal measurement that could not be included in the present volume. More down to earth is the magnificent tome by Horowitz and Hill (1989) on the art of electronics, with its many hints and details. The summary by Abernethy (1973) and the texts by de Sa (1981), Jones (1986), and Ott (1976) will also be found useful, not least for aspects of signal conditioning such as eliminating interference caused by ground loops. All these sources study frequency domain filtering and go on to describe phase-sensitive detectors – the topic of the following chapter. In addition, the author has long been impressed by the very readable book by Rosie (1966) on information and communication theory, which provides useful background on concepts such as bandwidth and noise.

13

Signal Recovery Using a Lock-in Amplifier

13.1 Introduction – the phase-sensitive detector

In this chapter we consider the synchronous detection approach to the recovery of signals from noise. In particular we assume that the signals are modulated or chopped at some convenient frequency, and that a reference waveform v_r is available at this frequency. A device known as a phase-sensitive detector (p.s.d.) is used to re-constitute the original signal, with the help of v_r.

The principle of operation of the p.s.d. is quite simple: a square wave generated from the input reference waveform is used to synchronously invert the incoming signal (see Figure 13.1). If the phase φ of the incoming signal is identical to that of the reference (for convenience we shall take the reference phase to be zero), the signal waveform is effectively full-wave rectified as in Figure 13.2(a), giving a maximum d.c. output. On the other hand, if the phases are 180° different, the full d.c. output is available but is inverted. In fact, the mean d.c. output level varies between these two extremes, and in general has the value $(2/\pi)v_s \cos \varphi$. Figure 13.2(b) shows what happens when $\varphi = 90°$, and Figure 13.2(c) shows the situation when $\varphi = 135°$. Clearly, it will be useful to have full control over the phase by means of a suitable circuit, and indeed the p.s.d. is only one component of the complete laboratory instrument required for synchronous detection, which is known as a lock-in amplifier or "lock-in". In fact, a lock-in amplifier (Figure 13.3) will include a signal amplifier, a reference section (amplifier, phase-shifter and limiter), a p.s.d., a signal averaging section or time-constant, an output d.c. amplifier, and a power supply (not shown in Figure 13.3).

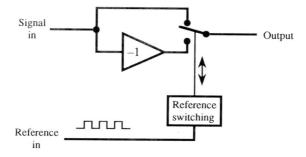

Fig. 13.1   Principle of operation of a phase-sensitive detector.

13.2   Circuits for phase-sensitive detection

Figure 13.4 shows the basis of a circuit for phase-sensitive (or synchronous) detection. On one half-cycle of v_r both diodes are reverse-biased. On the other half-cycle they are both forward-biased, and the two resistors pass currents determined respectively by $v_r + v_s$ and $v_r - v_s$. Since the output voltage v_o is equal to the difference between these two voltages, it is proportional to v_s and independent of v_r. In fact, this circuit is rather inefficient since it only provides an output current for half of the cycle of v_r. In a "double balanced modulator", such as that shown in Figure 13.5, this deficiency is avoided. Note that this type of circuit will operate even if the reference and signal inputs are interchanged. Note also that it is an a.c. circuit designed for use in the 1–500 MHz region (it normally employs low-loss ferrite-cored transformers).

If a circuit is required that will operate down to d.c., we must eliminate the transformers: one way of accomplishing this is to revert more closely to the form of circuit shown in Figure 13.1. A great many circuits have been designed for this purpose, but here we shall concentrate on just one of them. This circuit (Figure 13.6) uses an FET for performing the switching, and can be considered a special case of the operational amplifier circuit shown in Figure 13.7, where R_g can assume the values 0 and ∞. Analysis of the general circuit is quite simple, if the operational amplifier is ideal and takes negligible input current. In that case we find:

$$\frac{v_o}{v_s} = \frac{a_V(R_1 R_g - R_2 R_f)}{(R_2 + R_g)(R_f + a_V R_1)} \approx \frac{(R_1 R_g - R_2 R_f)}{R_1(R_2 + R_g)} \tag{13.1}$$

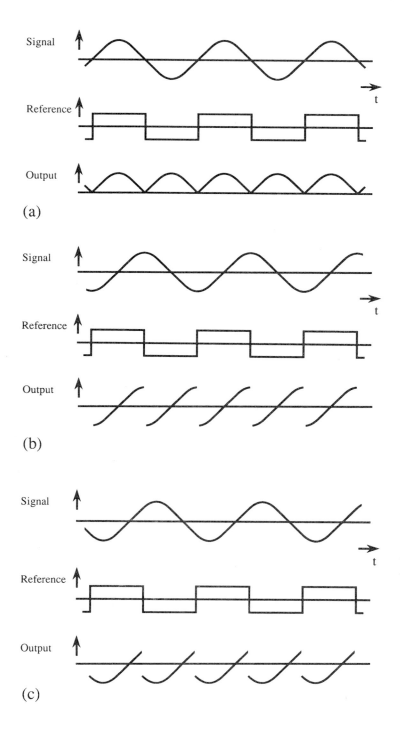

Signal

Reference

(a)

Signal

Reference

(b)

Signal

Reference

(c)

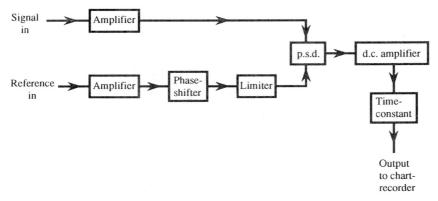

Fig. 13.3 Block diagram of a lock-in amplifier. This diagram shows how a lock-in amplifier is constructed using a p.s.d. as its central working unit. Unlike the p.s.d., the lock-in amplifier is a complete laboratory instrument: however, its power supply is not shown in the diagram.

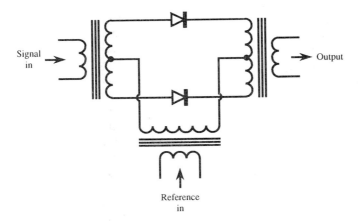

Fig. 13.4 Basis of a balanced modulator form of phase-sensitive detector.

Fig. 13.2 Action of a phase-sensitive detector on a sinewave. This figure shows the action of a phase-sensitive detector (p.s.d.) on a sinewave, (a) when there is zero phase-shift between the signal and reference waveforms, (b) when the phase-shift is 90°, and (c) when it is 135°. In (a) the p.s.d. acts essentially as a full-wave rectifier, but in other cases the mean amplitude is reduced and can go negative as in (c).

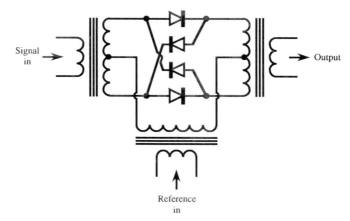

Fig. 13.5 Double balanced modulator form of phase-sensitive detector. Note that this is a completely passive type of circuit and has no power supply. This type of circuit is well-suited to operation in the 1–500 MHz region, and will not operate down to d.c.

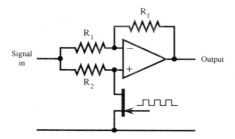

Fig. 13.6 Phase-sensitive detector capable of operating down to d.c.. In this circuit, it is intended that the FET will be switched continually between short and open circuit conditions by the reference waveform, thereby emulating the idealised circuit of Figure 13.1.

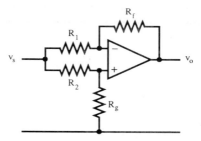

Fig. 13.7 Generalized model of p.s.d. circuit shown in Figure 13.6

where we have finally taken a_V to be large. If now we make $R_g = \infty$, we find:

$$(v_o/v_s)_\infty = 1 \qquad (13.2)$$

whereas if $R_g = 0$, we find:

$$(v_o/v_s)_0 = -R_f/R_1 \qquad (13.3)$$

Thus the circuit will make an ideal p.s.d. if $R_f = R_1$ and R_g alternates between 0 and ∞. However, for more realistic values of R_g, the value of R_f can be minutely adjusted to make

$$\frac{(v_o/v_s)_\infty}{(v_o/v_s)_0} = -1 \qquad (13.4)$$

Finally, it is useful to make $R_2 = R_1$ in order to approximately equalize the input currents at the operational amplifier input terminals. Values of R_1, R_2 and R_f of about $10\,k\Omega$ would lie suitably between (i.e. at about the geometric mean of) the on and off resistances of the FET, to optimize the intrinsic balance of the circuit.

13.3   Choice of operating frequency

We now consider the factors which govern the choice of operating frequency of a lock-in amplifier system. There are two main factors to be considered. One is that the frequency should be at a point where noise is fairly low, e.g. away from mains frequencies, and if possible at 1 kHz or higher to minimize flicker noise. (In general, we should set the modulation frequency above the flicker noise "corner" frequency f_c, as shown in Figure 13.8.) The second is that the frequency should not be in a position where the physical system undergoing investigation has a reduced response. For example, many physical systems take some time to come to equilibrium, so that too high an operating frequency would reduce the available signal. Quite often a compromise between these two factors is required, in order to optimize the SNR.

The lock-in amplifier time-constant can be increased to limit the output bandwidth and hence reduce noise. Of itself, this does not cut down the signal, since at this stage the signal has essentially been converted to d.c. However, to obtain *useful information* from the system, the d.c. level will be caused to change slowly, perhaps by altering the temperature or magnetic field in the system being investigated; this means that the signal is not quite

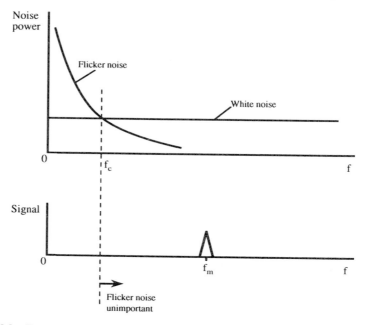

Fig. 13.8 Frequency diagram for a.c. detection system. This figure shows the frequency spectra of white noise and flicker noise. The operating frequency f_m of an a.c. detection system is deliberately chosen above the flicker noise "corner" frequency f_c. This condition applies whether a p.s.d. is used or not (as in the case of Figure 12.1).

a pure d.c. signal but contains very low-frequency a.c. components or sidebands (see Appendix D) – recall the well-known principle that the transmission of information requires a definite bandwidth. Clearly, increasing the time-constant reduces the rate at which information can be extracted, and then a limiting factor is the length of time available for the complete experiment to be performed. However, another limiting factor is how stable and free from drift the equipment is.

We can summarize the situation by noting that to increase the (voltage) SNR by a factor 10 necessitates an increase in observation time of $10^2 = 100$ (reasons for this squaring law are given in Section 14.2). For practical purposes, output time-constants of the order of $\frac{1}{2}$ minute are probably the longest that can be used – in particular, because drift in the output d.c. amplifier or pen-recorder will eventually obscure the signals (though this aspect might be less important if the whole process of synchronous detection were carried out digitally rather than using the analogue types of circuit envisaged so far).

13.4 A.c. detection of signal from a thermocouple

In this section we examine the detection of a signal from a thermocouple. Since a thermocouple takes time to stabilize to any new temperature, the electrical voltage from it can be expected to change approximately exponentially. In addition, when the thermocouple is subject to a strong source of heat, the rate of heating will have time-constant τ_1 which may be expected to be rather lower than the time-constant τ_2 for cooling (Figure 13.9). Thus two general equations govern periodic heating and cooling cycles:

$$S_h(t) = S_h(0) + (S_{max} - S_h(0))[1 - \exp(-t/\tau_1)] \qquad (13.5)$$

$$S_c(t) = S_c(0) + (S_{min} - S_c(0))[1 - \exp(-t/\tau_2)] \qquad (13.6)$$

If heating and cooling repeatedly take place for periods T_1, T_2, we have:

$$S_2 = S_1 + (S_{max} - S_1)[1 - \exp(-T_1/\tau_1)] \qquad (13.7)$$

$$S_1 = S_2 + (S_{min} - S_2)[1 - \exp(-T_2/\tau_2)] \qquad (13.8)$$

Eliminating S_1 we get:

$$S_2 - S_{min} = (S_{max} - S_{min}) \frac{1 - \exp(-T_1/\tau_1)}{1 - \exp[-(T_1/\tau_1 + T_2/\tau_2)]} \qquad (13.9)$$

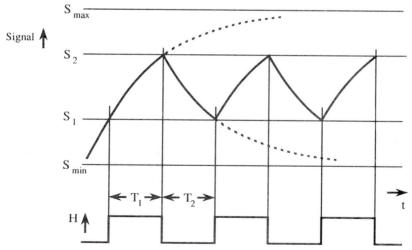

Fig. 13.9 Heating and cooling cycles for a thermocouple. This figure shows the repeated heating and cooling cycles when the a.c. voltage from a thermocouple is to be measured. H is the chopped input heating waveform.

To simplify the problem further we have to make some assumptions. Let us now take $\tau_1 = \tau_2 = \tau$ and $T_1 = T_2 = T$. Then we have:

$$S_2 - S_{min} = (S_{max} - S_{min}) \frac{1}{1 + \exp(-T/\tau)} \tag{13.10}$$

Similarly, by eliminating S_2 it is possible to show that:

$$S_1 - S_{min} = (S_{max} - S_{min}) \frac{\exp(-T/\tau)}{1 + \exp(-T/\tau)} \tag{13.11}$$

Hence the overall signal amplitude is:

$$\Delta S = S_2 - S_1 = (S_{max} - S_{min}) \frac{1 - \exp(-T/\tau)}{1 + \exp(-T/\tau)} = (S_{max} - S_{min}) \tanh(T/2\tau) \tag{13.12}$$

If we take the noise power as:

$$N^2 = \gamma/f = 2\gamma T \tag{13.13}$$

corresponding to flicker noise, we get an r.m.s. noise voltage:

$$N = \sqrt{2\gamma T} \tag{13.14}$$

Hence the voltage SNR is:

$$\rho = (S_{max} - S_{min}) \tanh(T/2\tau)/\sqrt{2\gamma T} \tag{13.15}$$

For $T \ll \tau$, $\rho \approx [(S_{max} - S_{min})/2\tau\sqrt{2\gamma}]\sqrt{T} \propto T^{1/2}$, so T should be increased, whereas for $T \gg \tau$, $\rho \approx [(S_{max} - S_{min})/\sqrt{2\gamma}]/\sqrt{T} \propto T^{-1/2}$, so T should be decreased. Clearly there is an optimum value of ρ for some intermediate value of T. Differentiating and setting $d\rho/dT = 0$ leads to the condition:

$$\sinh(T/\tau) = 2T/\tau \tag{13.16}$$

which results in the numerical solution $T/\tau \approx 2.18$. The maximum peak-to-peak signal is then:

$$S_2 - S_1 = 0.80(S_{max} - S_{min}) \tag{13.17}$$

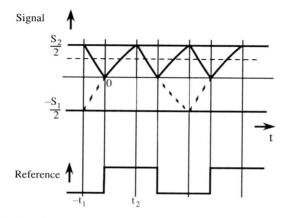

Fig. 13.10 Action of p.s.d. on thermocouple a.c. waveform. In this figure, the phase of the reference waveform is adjusted for optimal (i.e. full-wave rectified) output signal. The dotted lines show where the thermocouple signal has been inverted by the p.s.d..

and the optimum chopping frequency is hence:

$$f_{opt} = 1/2T = 1/4.36\tau \qquad (13.18)$$

Notice from Figure 13.9 that, in the steady state, there is a time delay between the heating source and the thermocouple output waveform: this will give a phase shift between the fundamentals of the two waveforms.

The calculation given above is useful since it gives the flavour of the optimizations involved when measuring physical phenomena from their a.c. response waveforms*. However, it makes use of an SNR that is derived from a *peak* signal. Although this may correspond to the SNR that the eye will observe on an oscilloscope screen, it is not the SNR that is relevant for a p.s.d. We now proceed to derive the SNR in the latter case. For convenience, we take the zero signal level as being on the symmetry axis of the waveform. In the vicinity of the new origin marked in Figure 13.10, the shape of the waveform before chopping in the p.s.d. is given by:

$$S = (S_0/2)[1 - \exp(-t/\tau)] \qquad -t_1 \leqslant t \leqslant t_2 \qquad (13.19)$$

---

*Indeed, it has now been shown that there is a frequency which optimizes the SNR, and this was the main point of this section. The following analysis (leading to equations 13.19–13.32) provides an extra degree of rigour which may be by-passed on first reading; in that case readers may jump to the last two paragraphs of the section.

where

$$S_0 = S_{max} - S_{min} \tag{13.20}$$

and after chopping it is given by:

$$S = -(S_0/2)[1-\exp(-t/\tau)] \qquad -t_1 \leqslant t \leqslant 0 \tag{13.21a}$$
$$S = (S_0/2)[1-\exp(-t/\tau)] \qquad 0 \leqslant t \leqslant t_2 \tag{13.21b}$$

The signal at the output of the lock-in amplifier is the mean of this waveform, and has the value:

$$\bar{S} = \frac{1}{T}\int_{-t_1}^{t_2} S\,dt = -\frac{S_0}{2T}\int_{-t_1}^{0}[1-\exp(-t/\tau)]\,dt + \frac{S_0}{2T}\int_{0}^{t_2}[1-\exp(-t/\tau)]\,dt$$
$$= (S_0/2T)\{(t_2-t_1)-2\tau+\tau[\exp(t_1/\tau)+\exp(-t_2/\tau)]\} \tag{13.22}$$

Returning to the definition of ΔS given in equation (13.12), we find:

$$\Delta S/2 = -(S_0/2)[1-\exp(t_1/\tau)] = (S_0/2)[1-\exp(-t_2/\tau)] \tag{13.23}$$
$$\therefore \quad \exp(t_1/\tau) = 1 + \Delta S/S_0 \tag{13.24}$$

and

$$\exp(-t_2/\tau) = 1 - \Delta S/S_0 \tag{13.25}$$

so

$$t_1/\tau = \ln(1 + \Delta S/S_0) \tag{13.26}$$

and

$$t_2/\tau = -\ln(1 - \Delta S/S_0) \tag{13.27}$$

We can now eliminate t_1 and t_2 from equation (13.22), eventually obtaining:

$$\bar{S} = (S_0/2T)(t_2-t_1) = (S_0\tau/T)\ln[\cosh(T/2\tau)] \tag{13.28}$$

In this case the optimization problem is slightly different. We again have:

$$N = \sqrt{2\gamma T} \tag{13.29}$$

$$\therefore \quad \rho = \overline{S}/N = (S_0\tau/\sqrt{2\gamma})T^{-3/2}\ln[\cosh(T/2\tau)] \tag{13.30}$$

ρ will be a maximum when $d\rho/dT = 0$, i.e. when

$$(T/2\tau)\tanh(T/2\tau) = (3/2)\ln[\cosh(T/2\tau)] \tag{13.31}$$

Solving this numerically gives $(T/2\tau) \approx 1.836$, so $T \approx 3.67\tau$, and the optimum chopping frequency is now:

$$f_{opt} = 1/2T = 1/7.344\tau \tag{13.32}$$

The reason that there is an optimum of this type is because the noise has been *modelled* as being entirely flicker noise. If the noise had been entirely white, it would have been better to go on increasing T indefinitely, thereby making the detected signal level tend towards to the peak signal level, while incurring no penalty from increased noise levels at low frequency. Unfortunately, flicker noise is often a fact of life, and its presence can clearly have a significant effect on the form of the final detection system: but in practice its effect will have to be minimized by employing a modulation frequency greater than the flicker noise corner frequency, as already noted (see Figure 13.8).

Finally, we should note that increasing T "indefinitely" as hinted above in the case of white noise would mean that the rate at which information could be extracted from the signals would be reduced to zero, so in practice a compromise is required – this reflects the fact that the above analysis was aimed *solely* at optimizing the SNR. It is also worth noting that the optimal SNR was calculated as a calculus maximum. Such maxima are relatively flat, so there will be little loss in SNR if the frequency deviates by up to $\sim 10\%$ of the value of f_{opt}.

13.5 Applying modulation frequencies with a lock-in amplifier

In Section 13.2 we described in broad terms how a lock-in amplifier can be used for optimal detection of signals in noise. In the case of the signal from a thermocouple, it turned out that the radiation producing the heating effect should be chopped at a rate equal to $1/7.344\tau$, where τ is the time-constant for exponential heating and cooling of the device; at this frequency, there is an optimal balance from the effects of flicker noise and signal averaging. This optimization is vital for maximizing sensitivity (SNR) in a situation where the sensory devices that are employed are rather slow in operation.

The experimental setup obtained by replacing the a.c. amplifier in Figure 12.1 with an a.c. amplifier and lock-in amplifier is not always the most straightforward to implement in practice. For example, it is common to superimpose a rapid magnetic field variation on a slow field sweep if an experiment involves drawing out variations in signal amplitude with magnetic field. In particular, magnetic resonance experiments are normally carried out in this way, as we shall see in the following section.

We end this section on a purely practical point. If a chopper wheel is used to modulate a signal as in Figure 12.1, a reference waveform can be obtained by a simple light-plus-photocell arrangement on a separate part of the wheel, the phase-shifter in the lock-in amplifer being used to eliminate any incidental time differences. (If a reference is not available or cannot be obtained in this way, a phase-locked loop will be required – see Chapter 18.)

13.6 Use of a lock-in amplifier in electron paramagnetic resonance

The phenomenon of electron paramagnetic resonance (EPR) may be explained from the energy level diagram of Figure 13.11. This shows that when certain types of magnetic ion (e.g. Fe^{3+} or Cr^{3+}) are placed in a crystal lattice such as aluminium oxide, their energy levels vary strongly with magnetic field. As a result, radiation of a certain frequency – often in the microwave 3 cm ("X-band") region – can induce transitions between the energy levels, and then radiation will be absorbed by the magnetic ions. It is quite difficult to employ microwave radiation over a wide range of frequencies with one apparatus (mainly because of the use of a high-Q tuned cavity – see below), and thus the microwaves are normally kept fixed in frequency while the magnetic field is swept over a suitable range (typically in the region 3–10 kG, 0.3–1 T*) – as indicated in Figure 13.11. A crystal detector can then monitor the level of microwave radiation reflected by the resonant cavity containing the sample, in order to detect any change in the radiation it absorbs. (A cavity resonator is used to concentrate the radiation on the sample, and gives an improvement in sensitivity by a factor proportional to the cavity Q, which may be as high as 5000.)

Figure 13.12 shows the basic form of the apparatus required for detecting EPR. In this apparatus, a directional coupler† is used to pass the radiation

---

*In common with much of the EPR literature, we denote magnetic field by H, but measure it in Tesla (the units of magnetic flux density $B = \mu_0 H$).

† This and other microwave devices mentioned in this section are described in more detail in Appendix C.

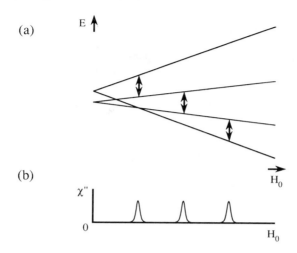

Fig. 13.11 Energy level diagram for a magnetic spin system. This figure shows (a) how the energy levels of a simple magnetic spin system vary with the applied field strength H_0. When an oscillating field of the right frequency is also applied, transitions (see the double-ended arrows) occur between the energy levels, and absorption of power can be detected. This effect is known as electron paramagnetic resonance (EPR). It is usually convenient to keep the frequency of oscillation (normally in the microwave region) constant, and to sweep the magnetic field H_0 in order to detect resonance lines, as in (b). The magnitude of the absorption is generally denoted by the susceptibility absorption coefficient χ''.

emerging from the cavity to the crystal detector. However, a 3 dB coupler causes a 3 dB loss of power from the source of microwaves (here a klystron) to the cavity, and a further 3 dB loss of power from the cavity to the detector: these losses do not occur with a circulator†. On the other hand, a magic tee† is sometimes used instead of either type of device, in spite of the loss of 3 dB it introduces into each of the two paths, since it permits a reference microwave signal of adjustable phase and attenuation to be fed to the crystal detector (see below).

Figure 13.12 also shows how the lock-in amplifier is used. As the magnetic field is swept slowly over the selected range, it is modulated rapidly at around 100 kHz by special small field coils near the sample, and the 100 kHz signal arriving at the crystal is detected by the lock-in. There is an interesting subtlety in the response of such an apparatus. As the magnetic field passes through the position of maximum response on the resonance, the modulation causes the χ'' absorption signal to be reduced on both sides of the maximum, and there is no oscillating response at the fundamental of the modulation

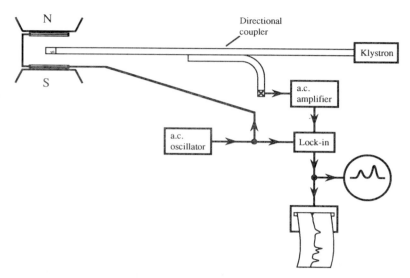

Fig. 13.12 Basic apparatus required for detecting EPR signals. Here microwave radiation from a klystron passes down a waveguide into a resonant cavity containing the magnetic sample. The EPR signal to be detected passes via a directional coupler to a crystal detector (the small box marked " × "), and the rectified signal proceeds through an a.c. amplifier to a lock-in. The reference to the lock-in has the same frequency as the modulation on the magnetic field H_0 at the sample.

Note that an additional wide field modulation at the supply frequency (not shown) can be used to display resonance lines on the oscilloscope: this arrangement is often used both for preliminary adjustments to the equipment and for rough measurements.

frequency (100 kHz) – though there is a small response at *double* this frequency (see Figure 13.13). Thus the response of the lock-in amplifier passes through zero at the centre of an absorption signal. If the amplitude of the modulation field is small, this effect will clearly produce a derivative of the original signal profile. On the other hand, making it too small will reduce the amplitude of the derivative signal, thereby reducing sensitivity. Clearly, there is a tradeoff between sensitivity and fidelity to the true derivative signal.

A rather intriguing method is used to optimize the signal in EPR systems. This method relies on the fact that there are actually two components to the magnetic signal – the χ'' absorption signal and the χ' dispersion signal (Figure 13.13). Now the derivative dispersion signal has a clearly defined maximum value at its centre. Hence, by adjusting the attenuator and phase-shifter in a reference arm attached to a magic tee (Figure 13.14), the derivative dispersion signal can conveniently be maximized for a suitable

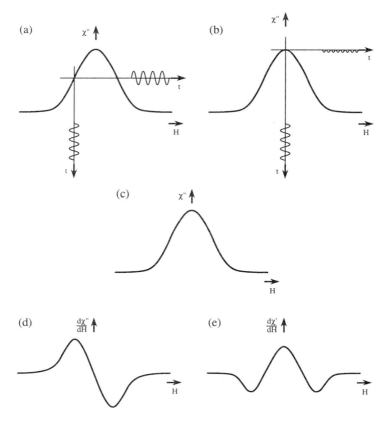

Fig. 13.13 Detection of absorption and dispersion signals using a p.s.d. Here (a) shows how modulation affects an absorption signal χ'', while (b) shows what happens at the centre of an absorption line – namely that there is only a small response at *double* the modulation frequency. As a result the *derivative* absorption signal is observed, as in (d). It turns out that, in some conditions, a dispersion signal χ' is detected instead of χ'' (see text). χ' appears very like the derivative absorption signal shown in (d), but when detected by a p.s.d. it appears as shown in (e).

large resonance; finally, the absorption signal can be obtained by adjusting the phase (only) of the microwave reference arm to give zero at the centre of the strong resonance (the derivative absorption signal passes through zero exactly on resonance). This setup procedure depends upon the existence of a non-absorbing variable phase-shift device that can be inserted in a microwave waveguide. It also has the advantage that the microwave detector crystal can be biased so that it will operate in its linear instead of its

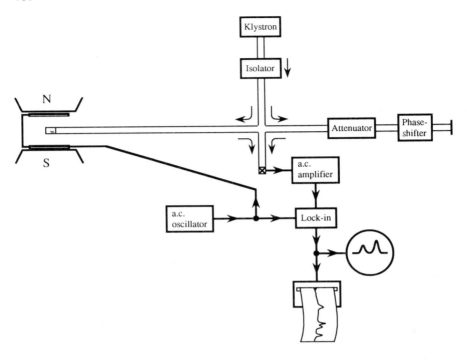

Fig. 13.14 Modified EPR apparatus capable of detecting absorption and dispersion signals. This modified version of the apparatus shown in Figure 13.12 includes the special microwave devices required for selecting the phase of the microwave signal to be detected, and thereby can detect the χ'' *or* the χ' signal. The wide field modulation and oscilloscope arrangement of Figure 13.12 is often helpful for setting up on the χ'' or χ' signals.

square-law region*. (Notice that the reason that a reference wave can be made to pick out an absorption or dispersion signal at will is that the reference wave is much stronger than either signal, and hence the amplitude of the resultant wave is approximately equal to the reference amplitude plus that of the signal having the same phase as the reference – see Figure 13.15.)

The procedure described above for optimizing the EPR signal is most easily accomplished when the signal is visible on an oscilloscope screen. For

---

*Note that with the magic tee arrangement, the crystal detector does not receive any *direct* radiation from the klystron, so the reference arm is able to provide a controlled source of radiation to the detector.

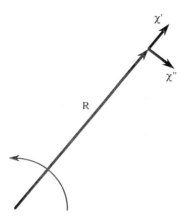

Fig. 13.15 Selection of absorption or dispersion signal. Here R represents the reference phasor, and controls the phase of the signal detected at the microwave crystal. In this case the χ' signal adds coherently to the constant reference amplitude, while the orthogonal χ'' signal is ignored to a first approximation.

this purpose it is necessary to have a low-frequency field modulation in addition to the modulation at 100 kHz: Figure 13.14 shows how this is achieved using 50 Hz sinusoidal modulation. Note that the 50 Hz signal cannot be detected easily with the time-constants of 0.1 to 1 second that are normally used at the output of the lock-in amplifier, and a time-constant in the millisecond range is employed in this special case – with the result that the SNR is relatively poor.

Finally, we note that the choice of the modulation frequency (assumed above to be 100 kHz) is the result of several factors, including the minimization of flicker noise and the possibility of designing high-frequency, low-noise lock-in amplifiers. In fact, 100 kHz has become a fairly standard modulation frequency in EPR. A further way of cutting down flicker noise is to use a superheterodyne detector operating at an intermediate frequency of 30–60 MHz. This reduces overall noise power by a further factor of 100 or so, and results in the *minimum* number of electron spins that can be detected being reduced by a factor ~ 10. When a superhet is being used, it is important to minimize the local oscillator noise. This can conveniently be achieved by use of a balanced mixer constructed from a magic tee and two microwave crystals – the latter being connected so as to cancel the local oscillator noise but to retain the full signal from the microwave cavity (Figure 13.16). This is possible since the two arms of the magic tee containing the detector crystals are fed in-phase by the cavity arm and in anti-phase by the local oscillator arm.

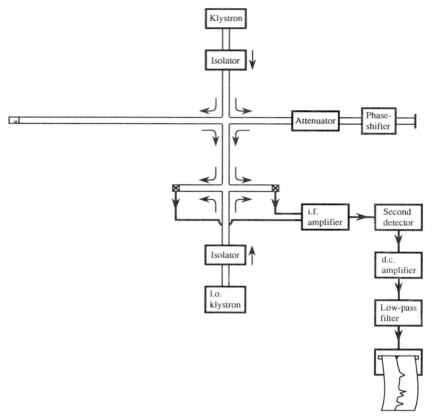

Fig. 13.16 Superheterodyne EPR system. This apparatus is a modified form of that shown in Figure 13.15, but incorporates superheterodyne detection. Thus there is a local oscillator klystron, and a balanced mixer employing two microwave mixer diodes and an additional magic tee. The use of a balanced mixer is necessary to minimise local oscillator noise. For clarity, the magnet and modulation coils are omitted from this diagram.

13.7   Electron nuclear double resonance

When one is interested in the sites near to magnetic ions in a crystal lattice, it is natural to study the magnetic resonances of nuclei in these regions, since the resonances undergo small shifts which can give information on the individual nuclear environments. As a topic, nuclear magnetic resonance (NMR) is similar to EPR, except that the magnetic moments of nuclei are typically ~ 1000 times smaller and more difficult to detect than those of electron spins. However, by "sitting on" (i.e. observing) an electron resonance it is possible to enhance the detection of the resonances of the nuclei that are near to, and strongly coupled with, the electron spins. To achieve this we just monitor the electron resonance signal, and subject the crystal sample to an oscillating "nuclear" magnetic field $H_N \cos(2\pi f_N t)$, where f_N is slowly swept through the nuclear resonance frequencies (typically in the range 5–30 MHz). This technique is called ENDOR (Electron Nuclear DOuble Resonance).

Technically, it might be possible to perform ENDOR by modulating the nuclear magnetic field H_N without modulating the electron resonance field H_0. However, for reasons that cannot be discussed in detail here*, H_0 modulation is normally retained. Thus the equipment employs two lock-in amplifiers, one fed directly from the other. The nuclear magnetic field is generally frequency- rather than amplitude-modulated, in order to cut down effects such as those due to variations in the amplitude of H_N and consequent variations in temperature of the sample crystal. (In this context, note that many of these physics experiments are carried out at low temperatures, so that the resonance lines will not be broadened so much by atomic vibrations.)

The resulting apparatus is shown in Figure 13.17. One of the main problems of ENDOR is that the signals are so small that they cannot normally be observed above noise on an oscilloscope screen and then it is difficult to optimize the system, e.g. with regard to nuclear modulation frequency. Overall, the physical systems that are studied by ENDOR can be very complex, with the result that it is not clear how to design an optimal detection system. For example, the range of frequencies being handled in the apparatus of Figure 13.17 includes $\sim 10^{10}$ Hz for the electron resonance signal, ~ 10 MHz for the nuclear resonance signal, ~ 100 kHz for the electron resonance modulation, ~ 30 Hz for the nuclear resonance modulation, with an option of a 50 Hz "slow" field modulation for oscilloscope presentation. Spectroscopy is then difficult for two reasons: not only must the correct range of frequencies be swept through – and this range of frequencies is not fully known until *after the resonances have been found* – but also it cannot be known fully in advance what experimental conditions (such as amplitudes and modulation frequencies) will give the best SNR. These statements apply with some force to ENDOR, though they also apply in other areas of spectroscopy. Broadly speaking, this situation does not apply in radar and other engineering systems (though the presence of various types of interfering

---

*In EPR, the modulation of H_0 at a frequency in the 100 kHz range causes "fast passage" effects which enhance intrinsic EPR sensitivity as well as permitting use of a phase-sensitive detection system.

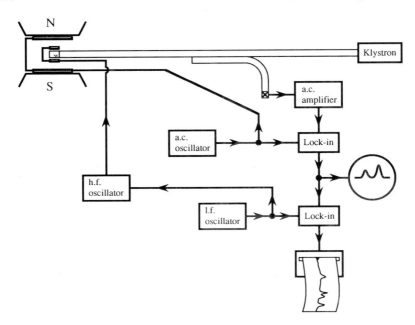

Fig. 13.17 ENDOR detection system. This is a modified version of the EPR system shown in Figure 13.12. An additional "nuclear" frequency is applied to the sample to stimulate nuclear resonances, and these ENDOR signals are detected as variations in the EPR signal. To optimize sensitivity, a further lock-in amplifier is employed, and this requires the nuclear frequency to be amplitude or frequency modulated at a low frequency (see text).

signals or "clutter" gives an alternative type of complication – the problem of lack of knowledge of the type of signal is then replaced by lack of knowledge of the type of noise!). Typical ENDOR spectra are shown in Figure 13.18; in the case of Figure 13.18(b), the SNR is rather small and some difficulty arose in getting large enough signals for accurate measurement.

Finally, note that although this type of spectroscopy may at first sound rather exotic, in fact it has achieved wide use. Furthermore, double resonance has been applied in other areas, for example, to detect EPR by its effect on *optical* resonances. The experimental details naturally change between these different areas of application, yet the apparatus used shows great similarities (e.g. in use of lock-in amplifiers) to that described above. And of course, double resonance is an approach that is necessarily particularly well suited for sensitively detecting and isolating specific signals from coupled systems, which might otherwise be submerged in a plethora of other confusing signals.

(a)

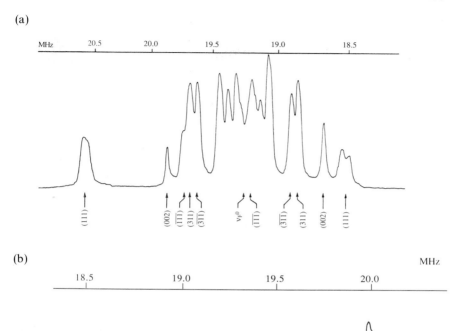

(b)

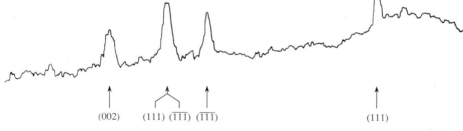

Fig. 13.18 ENDOR spectra for Ce^{3+} in CaF_2. This figure shows ENDOR spectra for Ce^{3+} (substituted for Ca^{2+}) at tetragonal sites in calcium fluoride. The resonances are from ^{19}F spin-$\frac{1}{2}$ nuclei; these would normally have resonances (v_F^0) around 19.3 and 12.8 MHz in (a) and (b) respectively, in the particular magnetic fields used in the two cases. Here, the resonances are spread widely by the Ce^{3+} magnetic fields at the various nuclei, depending on their positions (see numbers in brackets) relative to the Ce^{3+} ions (Baker *et al.*, 1968). In (a) $\mathbf{H_0}$ is parallel to (100); in (b) $\mathbf{H_0}$ is parallel to (111) (only part of this spectrum in shown). (a) and (b) were valuable in uniquely identifying the interstitial F^- ion at (002). Notice the high SNR in (a) and the fortuitously low SNR in (b).

13.8   Astronomical application of the p.s.d.

It would be wrong if this chapter finished without mentioning the use of the p.s.d. in radio astronomy where it was first developed. Radio astronomy involves the reception and detection of tiny signals from stellar objects. So tiny are many of the signals that large steerable aerials have to be employed before there is any chance of detecting them: witness the famous 250-foot radio telescope at Jodrell Bank. However, even with the enormous gain from such antennae, astronomical signals can still be exceedingly difficult to detect. For this reason Dicke invented a special radiometer, which involved comparing the intensity of the radiation arriving at the receiver with a signal from a calibration source. The comparison was performed by a switching device which looked alternately at the two signals and gave an output proportional to the difference between them. Dicke used a chopping frequency of 30 Hz and the output was observed with a d.c. meter giving an effective time-constant of about 1 second. His 1946 paper shows all the characteristics of a p.s.d. circuit (and many of those of a modern lock-in amplifier), though there are some details in which his experiments differ from those already described in this chapter. First, the artificial signal with which the incoming signal is compared is pure black-body radiation, i.e. thermal noise; in this it matches the characteristics of the incoming signal, and this arrangement is known to optimize detection, provided that the two signals also have closely the same mean square amplitude. Second, the incoming signal is only looked at half the time, which means that half the signal is wasted and the sensitivity is a factor $\sqrt{2}$ lower than it would ideally be; however, this problem could be eliminated if two synchronized choppers were used and the outputs summed.

Radio astronomy still uses p.s.d.s with long integration times. However, more modern forms of Dicke's experiment involve chopping between two antennae looking at orthogonal directions in space. Dicke's methods have particular application when measuring the radiation from the "big bang". The NASA Cosmic Background Explorer (COBE) satellite has recently (early 1992) been measuring the temperature of the cosmic background radiation: preliminary results give the temperature as 2.735 K with local variations of $30\,\mu$K corresponding to the fluctuations required 300,000 years after the big bang to produce clusters of galaxies. In the attempt to achieve this precision, three differential detectors looking at three separate short radio wavelengths – 3.3, 5.7 and 9.5 mm – had to be used, and the resulting 70 million measurements analysed in detail by computer (Henbest, 1992). Though it is too soon to give a full scientific appraisal of this work, these details are given to show that p.s.d.s have their origin in radio astronomy and still have a place there, but with the enormous data throughputs now required, it is often necessary to combine traditional (analogue chopping) and modern (digital computer) methodologies.

13.9 Summary

This chapter has studied the synchronous detection approach to the recovery of signals from noise. The phase-sensitive detector (p.s.d.) is the basic device that performs this task, and it is incorporated into the complete laboratory instrument known as the lock-in amplifier or "lock-in". A number of p.s.d. circuits exist, but fundamentally a p.s.d. need be no more complicated than an operational amplifier which is switched between inverting and non-inverting modes by an FET, or a 4-diode bridge with a set of coupling transformers. The other vital ingredient for a lock-in is some form of temporal averaging device, such as a time-constant τ: the latter will normally be in the range 0.1–1 s, with values of 10–20 s being considered the maximum practical for a purely analogue system, because long time-constants accentuate drift and noise from the final d.c. amplifier. For further details of modern lock-in amplifiers, see Section 19.2.3.

Choice of operating frequency is governed mainly by the need to exceed the flicker noise corner frequency. However, setting the output bandwidth is more complex, and devolves into a tradeoff between the need to suppress noise and the need to obtain information from signals at a reasonable rate without distorting them.

In some physical experiments, chopping the input radiation is impracticable and instead the *frequency* of the input radiation (or some other relevant parameter at the sample being studied) is modulated, and the resulting variation is monitored with a lock-in amplifier. Thus, in EPR, the magnetic field applied to the sample is modulated and a lock-in is used to monitor the changes in microwave radiation absorbed by the sample. The later sections of the chapter take various magnetic resonance experiments as case studies, so as to illustrate relevant principles of signal recovery in some depth. (See also Chapter 17 for pulsed magnetic resonance techniques.)

13.10 Bibliography

As remarked in Chapter 12, the volumes by Robinson (1974), Wilmshurst (1985) and Horowitz and Hill (1989) all study phase-sensitive detection. However, the second of these titles includes very few details of practical applications such as magnetic resonance, whereas Robinson's book has a useful amount of material on this topic. The third title does not dwell on any application area, but has a wealth of practical detail on all sorts of applications, and complements the present volume admirably. In addition, Kitchen (1984) contains information on the application of p.s.d.s in a variety of areas of astrophysics. There are a number of specialist books on magnetic resonance: these include the highly readable summary by Orton (1968), the impressive volume on nuclear magnetism by Abragam (1961), and the (once) comprehensive treatise on EPR by Abragam and Bleaney (1970). The latter also discusses ENDOR, but for details of experimental methods on ENDOR the reader is referred to the paper by Davies and Hurrell (1968) and the review in Baker, Davies and Reddy (1972). Note also the optical double resonance work reviewed (1976) by Davies (J.J. *not* E.R.) which confirms that double resonance is a general technique rather than a once-off idea. Finally, Dicke (1946) should be credited with the invention of the p.s.d., while the idea for the FET-switched p.s.d. circuit (Section 13.2) is due to Clayton (1973); Meade (1983) provides a fairly recent review of the p.s.d. and its applications.

13.11   Problems

1. Give a full derivation of equation (13.1). Investigate whether it is optimal to make R_1, R_2 and R_f about equal to the geometric mean of the FET on and off resistances as suggested at the end of Section 13.2.
2. Give a full derivation of equation (13.28.)
3. Give a modified version of equation (13.29) that includes flicker noise, with a flicker noise corner frequency of f_c. Hence derive more general equations than equations (13.30–13.31), and deduce the optimal chopping frequency when (a) $\tau = 1/4f_c$, (b) $\tau = 1/f_c$, (c) $\tau = 4/f_c$. What justification is there for generally taking the optimum chopping frequency as f_c?
4. Show that when a smoothly varying symmetrical wave such as a sine-wave is being detected using a p.s.d., the output is, to first order, independent of small errors in setting the phase of the reference.

14

Signal Averaging Techniques

14.1   Introduction

We have already seen how noise can be reduced by filtering in the frequency domain, and have found that the lock-in amplifier is extremely adept at this task. Here we examine an alternative approach: filtering in the time domain. The principle is simple. If no signal should appear during a given interval of time, we prevent noise from this interval from reaching the measuring device. Interestingly, although the lock-in amplifier acts as an extremely effective filter in the *frequency* domain, it actually operates by a process of switching in the *time* domain, and in many respects the methods we shall meet below are obvious extensions of those we have already covered.

14.2   Signal recovery using a boxcar detector

The boxcar detector is a device that has the capability for improving SNR by adding repetitive signals. It also has a temporal gating circuit for excluding noise at instants when no signal is expected (Figure 14.1). Thus there are two main requirements for a boxcar detector: (1) the gate should be open only long enough to include all the signal; (2) the gate should be opened every time a signal is due to appear. However, requirement (2) raises the question of when a signal should be made to appear, since physical systems normally emit signals only in response to stimuli. In fact, the system should be stimulated to emit signals as often as possible*. This is because, the more

---

* However, it is important to make the proviso that many physical systems give a lower-amplitude signal if they are stimulated too often, since some recovery period is often required. This has already been seen in Section 13.4 for the case of thermocouple signal detection.

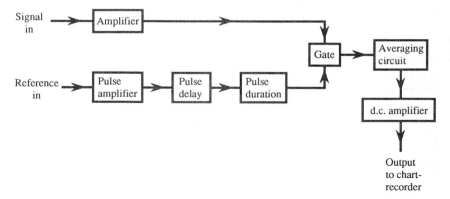

Fig. 14.1 Block diagram of a boxcar detector.

signals that come along in a given time, the more of them can be averaged to cut down the noise. To understand this quantitatively, note that p equal signals S_i sum to give a total signal voltage:

$$S = pS_i \qquad (14.1)$$

whereas the noise *powers* add:

$$N^2 = pN_i{}^2 \qquad (14.2)$$

giving an r.m.s. noise voltage:

$$N = \sqrt{p}N_i \qquad (14.3)$$

and therefore the voltage SNR is:

$$\rho = S/N = \sqrt{p}(S_i/N_i) \qquad (14.4)$$

Thus the power SNR is improved by a factor p – though if seen on an oscilloscope, for example, where one is really interested in the *voltage* SNR ρ, the improvement is only a factor $\sqrt{p}$.

Notice that there is no particular difficulty in recovering a signal which is initially buried in noise. In that case we go on adding up more and more signals and of course more and more noise, the signal voltages being summed and augmented coherently and the noise voltages adding incoherently: thus the SNR will eventually be greater than unity, i.e. the signal will emerge from below the noise. Of course this depends on the signal being in a known time-slot, so that it can be made to add coherently.

Next, we make an observation about nomenclature. The theory given above (equations 14.1–14.4) suggests that the SNR is increased by *summing* signals and noise powers. This means that after p signals have been added, the signal level has been multiplied by p. On the other hand, we also mentioned signal *averaging* to cut down the noise. Strictly, signal averaging only occurs when the summed signals are eventually divided by p. An *R-C* circuit (time-constant) carries out this normalizing operation implicitly – though in a digital circuit or digital computer it would have to be implemented explicitly. However, it is in the nature of the subject to refer to both processes as signal averaging, since the enhancement of SNR is achieved in either case.

So far we have assumed that we are merely averaging signals in order to improve the SNR. However, there is little point in having a device that gives high SNR if this is achieved by such a long output averaging time that the signal cannot be varied: the problem is that no information could then be withdrawn from the physical system under investigation. Clearly, we have to consider how the boxcar detector will be used. In practice, changes in the signal are induced by slowly sweeping some parameter ξ (such as magnetic field or temperature) in the system under investigation: these variations in signal are then recorded and give relevant information about the system. For an example of such a system see Section 17.3.

We now see, when setting up the boxcar detector, that the output time-constant should be increased to improve the SNR, but not to such an extent that the signal itself is distorted or reduced in size. As in the case of the lock-in amplifier, the output bandwidth should be sufficient to cope with any expected changes in the signal. In fact, there is a tradeoff between SNR and the rate at which information can be extracted from the system. In addition, a low initial SNR means longer observation time to obtain the same information about the physical system.

14.3 Computer averaging techniques

Several times in the past few sections we have seen that averaging of data permits better SNRs to be achieved. In particular, the boxcar detector approach led to the enhancement of the SNR in a specific time-segment if detection could be performed repetitively. This principle can evidently be extended by chopping long periods of time into short segments, each of which is handled by its own boxcar detection system. Clearly, this approach is potentially cumbersome, since boxcar detectors are quite complex instruments and it would require many of them to average a whole waveform. Nevertheless, a set of boxcar detectors would have several functions and overhead units in common, so there are obvious ways in which savings can be made*, and it should be possible to build a complete waveform averager on this basis.

It is clear that the approach indicated above – of having a complete analogue waveform averager constructed as a multiple boxcar detector – though practicable, would be excessively expensive. Such instruments were actually built in the 1960s, but at that period digital technology was advancing rapidly, and various companies were starting to market rather expensive but more general and powerful digital instruments that aimed to do the same thing more efficiently and with greater accuracy. In particular, such digital instruments could hope to have better uniformity between the different time-slots, so they would be more stable and more accurate. These early digital forms of signal averager were called by names such as CATs (computers of average transients), Enhancetrons, and so on. Many were not full-blown computers, but just dedicated pieces of digital hardware. However, during this period in the late 1960s, mini-computers were also becoming available, and these could be used for the same purposes if the correct data-collection front-ends (including analogue-to-digital converters) and interfaces were added to them. Soon, the CATs and other devices lost out to the digital computers, which were more general-purpose (though special-purpose devices continued to be available). Indeed, it is the general-purpose nature of computers and the almost infinite variation in the algorithms that they can run that is their permanent attraction for the purpose of data analysis and signal recovery.

This exclusivity and generality of the digital computer might itself appear to invalidate much of the work on signal recovery that has been covered in the preceding sections. There are several reasons why this is not so, or why this conclusion should be carefully qualified:

1. Just because a new technology emerges is no reason to make it displace all existing technologies and to abandon their principles of operation. Often the principles are unchanged, and the implementation alone is modified. This text concentrates especially on principles, and necessarily dwells on some of the simpler methods of implementation.
2. The problem of whether to use a computer implementation for a simple process or algorithm can be a difficult one: employing a computer implies a certain level of complexity – and the use of an interface that may

---

* For instance, a common power supply unit and a general timing unit could be employed, and these would give a substantial overall saving.

itself be more complex than the overall process to be carried out. Hence it may not be worth using a computer.

3. A computer may not give a better implementation, especially if very fast hardware is required, e.g. for fast averaging, or if low-pass filtering is to be carried out between two existing analogue processes. (In other words, why waste effort converting from analogue to digital format and vice versa if the task can be carried out efficiently and accurately without doing so?)

4. Use of a computer implies the capability for convenient enhancement of a process and for rather general data collection that might facilitate other approaches to data analysis, or (for example) automatic means of optimizing signal recovery. On balance, today we have to have sound reasons for *not* using the digital computer in any instance. However, in this book, it is nevertheless convenient at least to illustrate principles by studying the use of such devices as boxcar detectors and p.s.d.s.

5. Finally, note that certain analogue devices that could rival digital computers are coming to the fore, and will have to be considered increasingly in the future. Such devices include surface-acoustic wave devices, charge-coupled devices, optical computers, and neural networks – several of which have intrinsically analogue operation.

In what follows we consider some aspects of signal recovery for which the more detailed and flexible control of processes afforded by computer algorithms is helpful. Two particular examples are the shaped response of a signal averager to the last n samples, and the use of special algorithms for outlier rejection during signal averaging. We shall now study each of these approaches in turn.

14.4   The shaped response of a signal averager

Let us imagine that a simple R-C low-pass filter is used to average a signal. If it is fed from a low-impedance voltage source, a signal voltage sample at time t will gradually decay away with time-constant $\tau = RC$, while later voltage samples will also decay away with time constant τ, but from later moments corresponding to their inception (Figure 14.2). Thus we obtain an output signal $\tilde{S}(t)$ which is the convolution of the incoming signal $S(t)$ with the relevant impulse response $I(t)$*, where $I(t) = 0$ for $t < 0$ and $I(t) = (1/\tau)e^{-t/\tau}$

---

* The impulse response is defined as the response of the system to a very narrow (e.g. δ-function) pulse. For further details see Section 15.2.

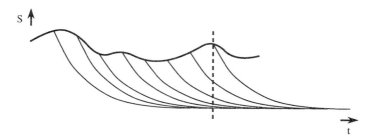

Fig. 14.2 Use of a simple R-C low-pass filter to average a signal. This figure shows how the instantaneous values on an input waveform are subject to exponential decay as they are integrated by a simple R-C low-pass filter. At the instant indicated by the dotted line, earlier signals have decayed to low level, while the current signal is at its maximum value. Future signals (which start after the dotted line) cannot contribute to the voltage at the selected instant. The result is the rather odd forgetting characteristic shown in Figure 14.3.

for $t \geqslant 0$:

i.e.

$$\tilde{S}(t) = \int_{-\infty}^{t} S(t_1)I(t-t_1)\,dt_1 \tag{14.5}$$

Specializing this to the case where $t=0$, we find that:

$$\tilde{S}(t) = \int_{-\infty}^{0} S(t_1)I(-t_1)\,dt_1 = \int_{-\infty}^{0} S(t_1)\frac{(\exp(t_1/\tau))}{\tau}\,dt_1 \tag{14.6}$$

We can interpret this equation as subjecting the signal function to a *forgetting* characteristic of time-constant τ and then averaging it – i.e. the forgetting characteristic is a rather odd form of weighting function (see Figure 14.3). It is clear that the characteristic is a forgetting one since earlier input voltages are necessarily *less* important at the output than the later ones, and all are less important than the current one.

The reverse exponential weighting characteristic arrived at above seems rather odd and has been arrived at by the accident of using a well-known circuit. We now ask what an ideal characteristic would be. A simple answer to this is that we might wish to weight all the last n samples *equally*, i.e. to employ a perfect moving-average filter with n equal weights, the samples

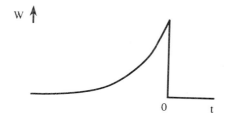

Fig. 14.3   Forgetting function for a simple *R-C* low-pass filter.

that are included in the weighting being the most recent *n* samples. This might be a difficult weighting function to institute by analogue circuitry, but it is rather easy to implement by dedicated hardware (e.g. a transversal filter – see Figure 14.4) and trivial to implement using a digital computer.

We can go further and consider other weighting functions, such as a Gaussian, which is also symmetrically centred around a particular instant, and derates in a reasonable way from this instant. It is not our purpose in this section to analyse possible weighting functions in any detail, but merely to note their existence and to emphasize that the digital computer provides an ideal means both of implementing them and of building a relevant database and trying a multitude of tests to determine which is best in any given situation.

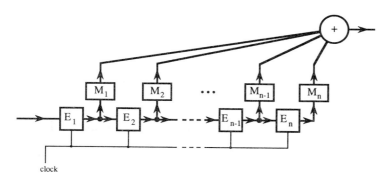

Fig. 14.4   Block diagram for a transversal filter. This is a clocked device, often constructed using CCDs: it is effectively an analogue form of the digital shift register. Charge held by the clocked elements E_i ($i = 1$ to n) is passed on at each clock pulse to the following element. Meanwhile, the various delayed outputs are taken through multiplier circuits M_i to an adder, which provides the output for the whole circuit. Such circuits are ideally suited to signal averaging and matched filtering.

Finally, note that *any* function we may select is bound to introduce a time delay: we can do little about this, as it is part of the nature of things. In the special case of a rectangular weighting function, the filter will cause a (minimum) time delay of half the time that it is averaged over, but beware of attempting to calculate the time delay for Gaussian or other weighting functions of infinite extent *unless* they are truncated first!

14.5 Outlier rejection in signal averaging

In the last section we considered various possible temporal weighting functions for signal averaging. In fact these were all linear weighting functions, corresponding to temporal convolutions with the input signal. Here we go further and examine whether linear weighting functions are necessarily the best ones to employ. In particular, suppose we have input voltage waveforms that are subject to impulse noise. Then some samples will be much larger or smaller than they should be (Figure 14.5). If we could determine which samples these are, then we might be able to arrange for them to be ignored, and the remaining samples could be weighted in some equitable manner to derive the final output waveform.

This type of strategy is well known in experimental physics, and in statistics, and is known as *outlier rejection*. The limiting case of outlier rejection is *median filtering*: in this case, outliers are repeatedly eliminated until only the central (median) sample remains (the *median* is the point in a distribution

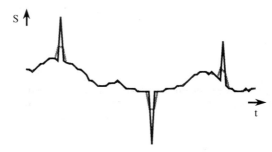

Fig. 14.5 Sampled analogue waveform exhibiting severe noise impulses. Here, the black line shows the original sampled analogue waveform exhibiting severe noise impulses. The grey curve shows the result of smoothing the original waveform using a 3-element moving-average filter. Note the blurring effect of the filter – especially in the region of the noise impulses.

Fig. 14.6 Smoothed version of sampled analogue waveform. This figure shows a smoothed version of the sampled analogue waveform with noise impulses already shown in Figure 14.5. In this case the smoothing is achieved by means of a 3-element median filter – i.e. the median (middle value) of each set of three consecutive intensity values is selected in turn and sent to the output. This cannot be achieved using a transversal filter, but a filter with the same basic clocked structure can be employed. Alternatively, a microprocessor or other digital computer can be used.

which divides the area of the distribution into two equal parts). Median filtering is widely used in the filtering of acoustic waveforms, cardiac voltage waveforms, and other similar 1-D signals; it is also extremely widely used in the processing of digital images. In all these cases it has the advantage of not blurring the incoming signal data (as does a normal averaging filter – see Figure 14.5), yet it suppresses noise and especially impulse noise very effectively (Figure 14.6). On the other hand, it is not so widely used in signal recovery applications, possibly because the early analogue means of averaging data could not easily incorporate the non-linear processing required for its implementation. It is only the advent of digital computers that has permitted such processes to be tested carefully and applied when this turns out to be useful.

Again, our primary purpose in the present section is not to advocate median or other non-linear filtering in signal recovery applications, but to point out the possibility of using them and to emphasize the value of the digital computer for (a) implementation and (b) testing of such functions. However, the author's experience of non-linear filtering in the area of image processing indicates the potential value of this technique in these other areas (i.e. physics experiments, radar and other specifically signal recovery applications). Indeed, non-linear filtering is only exemplified by median filtering, and the whole panoply of abstract reasoning and artificial intelligence can also to be applied to the process of extraction and recognition of signals and to the interpretation of data from physical sources. We defer further discussion of these points, and proceed instead to more detailed analysis of the extraction of signals from noise.

14.6   Signal averaging with different types of noise spectra

The advent of high-speed electronics and appropriate electronic (e.g. digital) storage systems has led to various possibilities for recovering low-level signals from noise. We here consider two paradigms. In the first we scan over the data-space slowly, taking multiple samples and averaging them as we go. In the second we scan repeatedly and rapidly over the data-space, and as we pass any given data vector we accumulate a signal into an appropriate channel, thereby averaging over the whole data-space many times. This latter approach can thus be modelled as doing a simultaneous scan over the whole data-space, and then another such scan, and then another, gradually building up and averaging all the signal channels simultaneously, thus progressively improving the SNR in each channel.

We now ask which of these two paradigms is the better one to use. First, we assume that the time-constants of the physical system are not at all disadvantaged by the rapid scans of the second paradigm: i.e. we assume that the signals are entirely unchanged on going from the one paradigm to the other. Thus the answer to our question must lie entirely in the nature of the noise.

Next we note one very simple result: if the noise spectrum is entirely white, then there is no *particular* advantage to be gained from using either method. Indeed, by specifying that the same number of samples go into each channel, and assuming that for each of these samples the same quantity of (mean square) noise is added, then there can be no relative gain for either method. Ultimately, the reason for this is that we are merely changing the *order* in which the noise samples are added, but not (statistically) their values; and with white noise, noise that appears in any temporal sample is statistically identical to that appearing in any other temporal sample. (Note that genuine white noise has a uniform power spectrum, and we can conclude, as in Section 10.6, that all samples of the noise waveform are independent and subject to the same probability distribution.)

But what if the noise is not white and does not have a uniform power spectrum? Various possibilities exist, but here we content ourselves by taking the case of flicker noise, which has a $1/f$ spectrum. We can envisage this as giving the signal a fairly serious drift: then it is better to spread this drift as uniformly as possible over all channels. Thus we scan rapidly over the data-space, adding any given drift voltage as equally as possible to all channels, and then repeating this with any new drift voltage that arises (of course the drift changes gradually and not in steps, but the principle is still to average and thereby equalize the drift voltage in each channel).

Most noise sources have a flicker or drift component, which merges at higher frequencies into normal white noise. Thus we have to take particular care to ensure that signal recovery is not disadvantaged by the flicker noise component. This means that, in general, it is better to scan rapidly and repeatedly through the data-space, gathering and averaging appropriate data. However, care must be taken to ensure that the scan rate is not so high that the signal is cut down significantly by this procedure. In general, there will be a tradeoff between noise limitation and signal maximization, and only detailed theoretical modelling or careful experimentation can determine the optimum operating point. Note that all these considerations arose earlier – albeit with different terminology – in Chapter 13 describing synchronous detection systems. Clearly, we are involved here with the same underlying processes, and the solutions will ultimately be subject to the same limitations.

14.7   Offset, drift and flicker noise

Let us suppose we are detecting a signal which appears as a bleep (such as a radar echo signal) on a fairly steady background. The bleep can be degraded in a number of ways, additive white noise being perhaps the most obvious. However, it can also be on a background level that has drifted up or down, or which has not been preset properly to zero. In the latter case, we say that there is a d.c. offset, and this has to be allowed for in detecting the signal.

Drift, on the other hand is by definition a varying background level, though it can often vary quite slowly. This can lead to problems, since drift and offset may be indistinguishable in the short term, and a preliminary setup procedure may later be found to have been carried out incorrectly. Drift is normally taken to be the result of temperature changes in electronic circuits and devices, and we might have the notion that rigorous attention to thermal stabilization would eliminate it (i.e. an output drift voltage would be a strict function of temperature). In practice such a notion is untestable because the very currents that make the circuits work cause heating effects which will vary from one part of the circuit to another, and there is no *unique* temperature that could characterize the equipment. Nevertheless, we are left with the idea that drift is a very low-frequency voltage or current variation, which is likely to take place not far from linearly with time.

By contrast, flicker noise has a $1/f$ frequency variation, which therefore increases as the frequency is reduced – perhaps giving the impression that it is much the same as drift. Oddly, there is the startling possibility that it becomes infinite if we integrate the noise power over all low frequencies. However, this eventuality cannot materialize as there is always some limitation on the observation time. *Suppose* we take any decade of frequencies and integrate the noise power over this range. Then we will naturally always get the same result. Now any experiment will require a certain amount of information and a certain resolution, so it is possible to say that "comparable" experiments require the same number of decades of frequency variation: thus there is a sense in which error due to flicker noise is independent of observation time t_{obs} (or accuracy cannot be improved by extending the measurement time). This contrasts with the case of drift where we have seen that the error voltage due to noise is roughly proportional to t_{obs}, and to the case of white noise where the error voltage (after signal averaging) is roughly proportional to $t_{obs}^{-1/2}$. These considerations show that drift can in principle be distinguished from flicker noise, though many workers do not make this distinction, but merely regard drift as very low-frequency noise.

Fortunately, drift is less serious than flicker noise at high frequencies, so modulating signals and using a p.s.d. is a good way of reducing its effects. However, it should not be overlooked that the output d.c. amplifier in a lock-in amplifier will itself drift, and this effect tends to limit the time-constants that can be used to seconds rather than minutes or hours.

14.8 Optimum gate width for a boxcar detector

When a boxcar detector is used to detect a bleep such as a radar echo of roughly Gaussian shape, at first we imagine that it is best to open the gate to cover the whole duration period of the signal. However, at the extremes of this period the incremental SNR is very low and is necessarily less than the average SNR for the whole signal: for small time durations at the extremes of the period very little signal is being added but a considerable amount of noise is being included, so on balance it will be best to reduce the gate width – at least marginally. We can go further with this idea to find when an optimal SNR will occur.

Now the total signal for gate width $t_2 - t_1$ is:

$$Q = \alpha \int_{t_1}^{t_2} S \, dt \tag{14.7}$$

and the total (white) noise power is:

$$N^2 = \gamma(t_2 - t_1) \tag{14.8}$$

so the voltage SNR is proportional to:

$$\rho = \int_{t_1}^{t_2} S \, dt / (t_2 - t_1)^{1/2} \tag{14.9}$$

For an optimum we want $d\rho/dt_1 = 0$ and $d\rho/dt_2 = 0$. Now

$$d\rho/dt_1 = -S_1(t_2-t_1)^{-1/2} + \tfrac{1}{2}(t_2-t_1)^{-3/2}\int_{t_1}^{t_2} S\,dt \qquad (14.10)$$

$\therefore \quad d\rho/dt_1 = 0$ for:

$$S_1 = \int_{t_1}^{t_2} S\,dt/2(t_2-t_1) \qquad (14.11)$$

Similarly:

$$S_2 = \int_{t_1}^{t_2} S\,dt/2(t_2-t_1) \qquad (14.12)$$

$$\therefore \quad \int_{t_1}^{t_2} S\,dt = 2S_1(t_2-t_1) = 2S_2(t_2-t_1) \qquad (14.13)$$

which shows (Figure 14.7) that the integrated signal over the interval t_2-t_1 must be made double the value obtained by integrating the extreme signal

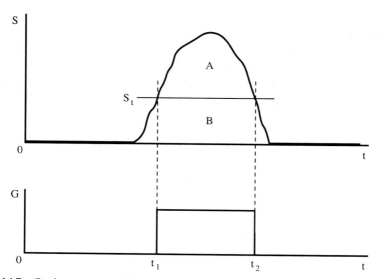

Fig. 14.7 Optimum gate width for a boxcar detector. This figure shows the threshold height S_t being adjusted so that the width of the time-slot over which the signal S is greater than S_t is such that the two areas A and B are equal. In that case the SNR is the best available with a boxcar detector under white noise conditions. G is the gate waveform.

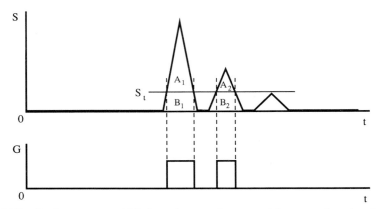

Fig. 14.8 Optimum gate width for a boxcar detector with a complex signal. Here the threshold height S_t is adjusted so that the total areas between the pairs of dotted lines are such that $\Sigma_i A_i = \Sigma_i B_i$. This optimizes SNR in this more complex case.

value over the same interval (or that the integrated signal in excess of the extreme signal value must be made equal to the integrated extreme signal value). Note that the analysis proves rigorously what is in retrospect obvious – that the two extreme signal values S_1 and S_2 have to be equal for optimum SNR.

This simple but somewhat curious result means that for a rectangular signal it will be best to retain the whole signal profile; for a triangular signal it will be best to retain only the central two thirds of the profile; and for a Gaussian signal, it will be best to retain the central ~ 3.87 standard deviations of the profile. Notice that the amount excluded is unrelated to the absolute noise level: indeed, the above calculation gives a result that is *independent of the input SNR*, which might well be much less than unity.

Finally, it may be noted that the above result may be generalized for any profile that is always positive or zero, whether symmetric or otherwise. We merely have to find a threshold height which segments the profile into time-slots, of which those with larger heights than the threshold integrate to give equal areas above and below the threshold (Figure 14.8).

14.9 Summary

This chapter has examined two main topics: the first is the use of time-slicing techniques in signal recovery, and resulted in extending the p.s.d. concept to linear gates and boxcar detectors; the second is signal averaging, and is a very powerful concept that transcends most other individual techniques in signal recovery. The truth of the latter statement is seen from the discussion of use of computers for signal averaging, which offers many possibilities that are not present with purely analogue approaches (as embodied in conventional boxcar detectors). However, signal averaging has also been seen to have its limitations, unless the concept is broadened to cover techniques such as median filtering which are capable of eliminating outliers, and "intelligent" processing which has the potential for identifying and eliminating interfering signal sources or clutter. It was also seen that signal averaging computers are able to scan repeatedly and rapidly over the data-space, thereby facilitating the elimination of drift and flicker noise. Of course, such methods require the physical system being studied to have a fast response time, so that the signals themselves are not attenuated by rapid sampling techniques. Finally, a means of estimating the optimal gate width for a boxcar detector was described: this forms a useful link with the work of the next chapter, on matched filtering.

No detailed case studies are provided within this chapter, as the concepts are extensively rehearsed and developed in Chapters 16 and 17 on radar and magnetic spin-echo systems.

14.10 Bibliography

As in the previous two chapters, it is useful to refer the reader to the volumes by Robinson (1974), Wilmshurst (1985) and Horowitz and Hill (1989). The last of the three is always the easiest to read (provided you know what to look for and restrict your scope appropriately – the number of distractions is enormous!), and has some interesting detail on topics such as Mössbauer spectroscopy and astronomy. On the other hand, the reader will find a more thorough theoretical development in Wilmshurst's book. Again, the latter unfortunately does not present much of the work in the form of case studies, therefore putting the volume more at the level of the serious practitioner than the undergraduate student. Abernethy (1970) provides a popular account of the boxcar detector. For further references on applications of boxcar detectors and related devices, see the bibliographies of later chapters, notably Chapters 16 and 17.

14.11 Problems

1. An aeroplane with a delta wing is flying directly towards a radar transmitter. Derive a formula showing how the overall SNR will vary

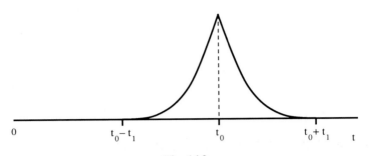

Fig. 14.9

when various portions of the signal (approximating to a right-angled triangle) are averaged by a linear gate. Deduce the condition for optimal SNR under these conditions. Confirm that the theory of Section 14.8 leads to the same result.

2. With the help of standard tables of Gaussian values, or otherwise, verify the statement made in Section 14.8 that a 1-D Gaussian signal is best detected by a boxcar detector when its gate is open over the central 3.87 standard deviations of the profile. Is the statement made in Section 14.8 about the optimal detection of a triangular signal exact or approximate? How does it depend on the shape of the triangle?

3. A signal may be approximated by a rectangular pulse of height h_1 and duration t_1, followed immediately by a rectangular pulse of height h_2 and duration t_2, where $h_1 > 2h_2$. Show that if t_2 is small, it is best to ignore the second pulse, and otherwise it may be best to average over *both* pulses. Find the conditions under which it is best to average over both pulses.

4. A certain physical system is stimulated by a short pulse of microwaves at time $t=0$, and a positive cusp-shaped signal emerges between times $t_0 \pm t_1$, with its peak at time $t=t_0$. The cusp is in the form of two parabolas back-to-back, which are respectively zero at the start and end of the signal (Figure 14.9). The peak height is known to be inversely proportional to the fourth root of the repetition rate of the initial microwave pulse. Determine how best to set up a boxcar detector to detect the signal. Obtain a numerical result if $t_0 = 5\,\mu s$ and $t_1 = 2\,\mu s$.

15

Matched Filtering Techniques

15.1 Introduction

This chapter covers a topic that is vital for the purpose of recovering low-level signals from a background of noise – matched filtering. The chapter is intended only to introduce the topic and to show how it may usefully be applied in one area – that of radar. However, its power goes well beyond the confines of radar, and it has implications for all the work covered in Part 3 of this book. We start in the next section by considering the detection of some very simple types of signal.

15.2 The concept and design of a matched filter

Consider a situation in which waveforms of the type shown in Figure 15.1 are fed into delay circuits and multipliers, and finally summed in an adder. If the delay-times are multiples of the basic unit of time for the waveforms, then the final output consists ideally of similar stepped waveforms.

We can easily find how the initial waveform is modified by examining the effect of passing a single narrow pulse through the system. The output of the circuit in that case is called its *impulse response*. Thus case 1 in Figure 15.2 shows the impulse response for the circuit of Figure 15.1; cases 2, 3, 4 illustrate the outputs for the same circuit when the input is not just a single, narrow pulse, while the following calculation shows how the output waveform

(a)

(b)

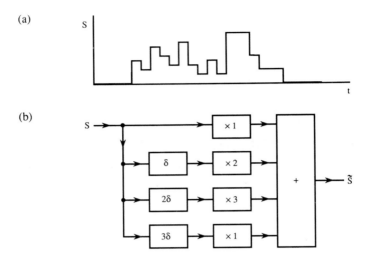

Fig. 15.1 Typical stepped waveform. (a) A typical stepped waveform which is an approximation to an initial continuously varying waveform. Stepped waveforms of this type result when continuous waveforms are sampled periodically. Part (b) shows, in block form, a circuit that might be used to process the waveform in (a): the various branches of the circuit delay the input waveform S by multiples of a basic delay time δ, and then multiply the signal by various factors before adding to give the output waveform $\tilde{S}$.

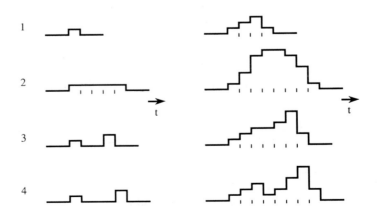

Fig. 15.2 Responses of the basic circuit to various input signals. This figure shows the responses (right) of the circuit of Figure 15.1(b) to various input waveforms (left). In case 1, a narrow input pulse gives an output which is defined as the *impulse response* of the circuit.

is constituted in case 2:

	1	1	1	1	1			
1	1	1	1	1	1			
2		2	2	2	2	2		
3			3	3	3	3	3	
1				1	1	1	1	1
	1	3	6	7	7	6	4	1

Note that it is also possible to calculate the output waveform by summing the impulse responses of the narrow pulses of which the input waveform is composed:

	1	2	3	1				
1	1	2	3	1				
1		1	2	3	1			
1			1	2	3	1		
1				1	2	3	1	
1					1	2	3	1
	1	3	6	7	7	6	4	1

This merely reflects the fact that the output waveform $\tilde{S}(t)$ is the convolution of the input waveform $S(t)$ with the impulse response $I(t)$ of the filter – a sum that can be written down in two ways:

$$\tilde{S}(t) = \sum_{t_1 = -\infty}^{\infty} S(t-t_1)I(t_1) = \sum_{t_2 = -\infty}^{\infty} S(t_2)I(t-t_2) \qquad (15.1)$$

Circuits of the type described above are clearly completely specified if $I(t)$ is known, so in many cases designing a circuit with particular properties reduces to defining an impulse response for it, and then deriving an appropriate circuit.

At this stage we identify a vitally important problem: how can we devise a circuit which will help to recognize a given type of signal $S(t)$ in the presence of interfering noise by optimizing its SNR? As an example, let us suppose we want to recognize signals of the type shown in Figure 15.3. Let us try various impulse responses for the processing circuit (the individual time-responses are illustrated in Figure 15.4):

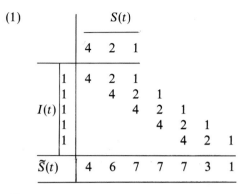

(1)

$S(t)$						
4	2	1				

$I(t)$							
1	4	2	1				
1		4	2	1			
1			4	2	1		
1				4	2	1	
1					4	2	1
$\tilde{S}(t)$	4	6	7	7	7	3	1

(2)

$S(t)$		
4	2	1

$I(t)$ 5	20	10	5
$\tilde{S}(t)$	20	10	5

(3)

$S(t)$		
4	2	1

$I(t)$				
4	16	8	4	
2		8	4	2	
1			4	2	1
$\tilde{S}(t)$	16	16	12	4	1

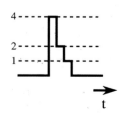

Fig. 15.3　Signal to be recognized.

(4)

		$S(t)$			
		4	2	1	
$I(t)$	1	4	2	1	
	2		8	4	2
	4			16	8 4
$\tilde{S}(t)$		4	10	21	10 4

(5)

		$S(t)$			
		4	2	1	
$I(t)$	1	4	2	1	
	1		4	2	1
	5			20	10 5
$\tilde{S}(t)$		4	6	23	11 5

(6)

		$S(t)$			
		4	2	1	
$I(t)$	2	8	4	2	
	2		8	4	2
	3			12	6 3
$\tilde{S}(t)$		8	12	18	8 3

As we are interested in the SNRs of the various filters, it is not sufficient to examine the values of $\tilde{S}(t)$: we must also work out the noise levels. Assuming a basic sample r.m.s. noise level of N_b, we obtain the following respective values for the r.m.s. noise:

(1) $N = N_b\sqrt{(1^2 + 1^2 + 1^2 + 1^2 + 1^2)} = N_b\sqrt{5}$
(2) $N = N_b\sqrt{25}$
(3) $N = N_b\sqrt{(4^2 + 2^2 + 1^2)} = N_b\sqrt{21}$
(4) $N = N_b\sqrt{(1^2 + 2^2 + 4^2)} = N_b\sqrt{21}$
(5) $N = N_b\sqrt{(1^2 + 1^2 + 5^2)} = N_b\sqrt{27}$
(6) $N = N_b\sqrt{(2^2 + 2^2 + 3^2)} = N_b\sqrt{17}$

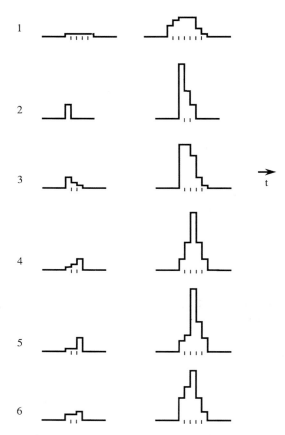

Fig. 15.4 Response with circuits of various impulse responses. Six circuits are defined by their impulse responses (left), and their responses to the signal of Figure 15.3 are indicated on the right.

In all cases, we have calculated the output noise level assuming that (a) (unlike the signal) it has a constant mean square value for all t, and (b) noise originating at different moments (and delayed by different amounts in the delay circuits before reaching the adder) is uncorrelated, so we have to sum the squares of the individual noise levels to obtain a final mean square noise level (see Section 10.5).

We can now deduce the respective peak (voltage) SNRs:

(1) $SNR = 7/\sqrt{5}N_b = 3.13/N_b$
(2) $SNR = 20/\sqrt{25}N_b = 4.00/N_b$

(3) $\mathrm{SNR} = 16/\sqrt{21}N_b = 3.49/N_b$
(4) $\mathrm{SNR} = 21/\sqrt{21}N_b = 4.58/N_b$
(5) $\mathrm{SNR} = 23/\sqrt{27}N_b = 4.43/N_b$
(6) $\mathrm{SNR} = 18/\sqrt{17}N_b = 4.37/N_b$

Of the six circuits defined by the impulse responses, we see that the fourth is the one giving the largest SNR. It turns out that this gives the highest SNR that is achievable by any such circuit for the signal of Figure 15.3. In fact the circuit which has the highest SNR for a given signal is called a "matched filter" and has an impulse response $I(t)$ which is the *time-reverse* of the signal $S(t)$ being sought. This is at first somewhat surprising – a priori one might have thought that $I(t)$ should have a form identical to that of $S(t)$ – but in fact the "reversed" impulse response is compensating for the asymmetry of the signal $S(t)$. There is a sound rationale for this: we are really aiming to get the whole signal to appear superimposed at one instant, to obtain the optimum SNR at that instant whatever happens to the output at other moments: to achieve this the leading parts of the signal must be suitably delayed and the trailing parts must be suitably advanced, while at the same time weighting the different parts of the signal in proportion to their expected amplitudes so as to preferentially suppress noise*. Clearly, this is a more sophisticated version of the gating system outlined in Section 14.7. Finally, note that when a signal is fed into a filter to which it is matched, the output waveform is symmetric, as in case (4) above: indeed, the symmetric output signal is then the autocorrelation function of the signal (this is easily proved by setting $I(t)=S(-t)$ in equation (15.1)).

The principles outlined above apply only in the case of background white noise, so that noise which has been delayed by elements in the matched filter is of an identical nature to undelayed noise, and is uncorrelated with it; only then is it valid to add the mean square noise levels, as above. If the noise is not white, then it is necessary to feed the input first through a "noise-whitening" filter and to use a modified matched filter to look for the new form of signal $S'(t)$ which emerges from the first filter.

15.2.1   Some implementation problems

In the above analysis we considered an initial waveform which was in all cases constant over intervals equal to some basic unit of time. This waveform is effectively a periodically sampled but continuously varying analogue signal.

---

*It will not be clear at this point why the weighting should be in *direct* proportion to the amplitudes, but in Section 15.5 we prove that this is the case.

It is evident that if we reduce the basic unit of time or sampling period, and increase the number of samples, we can obtain as good an approximation as we wish to any analogue waveform, including the signal $S(t)$ which we wish to recognize. It is also evident that a suitable matched filter – i.e. one having an impulse response $I(t) = S(-t)$ – can also be devised by keeping the basic circuit of the same form as described earlier, but increasing the number of delay and multiplication elements. Thus a matched filter suitable for any analogue signal $S(t)$ can in principle be built, though a practical version might be expensive because of the large number of circuit elements required*: such a situation might arise where either great accuracy in the signal processing circuitry is needed or the duration of the signal to be recognized is particularly long. On the other hand, a matched filter of the above type is relatively easy to construct when the signal is in the form of short pulses of known shape – as for example in certain radar applications.

In view of the expense of building an accurate matched filter of the above type, it is worth considering whether an alternative approach might be of use. This is indeed possible for some signal profiles. Consider the signal $S(t)$ shown in Figure 15.5(a), which has a curtailed rising exponential form. A suitable matched filter would have a decaying exponential impulse response (Figure 15.5(b)). This is readily realizable using a simple R-C circuit (Figure

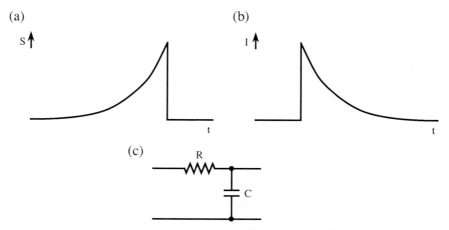

Fig. 15.5 A continuous signal and its matched filter. (a) A curtailed rising exponential signal profile; (b) the impulse response of the matched filter that should be used to detect it; (c) the simple circuit that is able to achieve this.

---

*However, a CCD "transversal filter" (based on an analogue shift register circuit) could fulfil many of these demands, but would have to be tailored into a special VLSI chip.

15.5(c)). However, signals of falling rather than rising exponential form are far more common: in this case we require a simple circuit with impulse response as in Figure 15.5(a), and in fact no purely analogue filter of this type is possible. This means that we must fall back on the type of matched filter discussed earlier.

15.3 Correlation

The correlation coefficient of two functions $x(t)$ and $y(t)$ is defined as:

$$R_{xy} = \frac{\displaystyle\int_{-\infty}^{\infty} x(t)\,y(t)\,dt}{\left[\displaystyle\int_{-\infty}^{\infty} x^2(t)\,dt \int_{-\infty}^{\infty} y^2(t)\,dt\right]^{1/2}} \qquad (15.2)$$

R_{xy} is zero if the two functions are totally uncorrelated, negative if they are negatively correlated, and positive if they are positively correlated: e.g. if $y(t) = \gamma x(t)$, for some constant γ, then $R_{xy} = 1$, while if $y(t) = -\gamma x(t)$, then $R_{xy} = -1$. On the other hand, if $x(t) = \sin \omega_1 t$, and $y(t) = \cos \omega_1 t$, then $R_{xy} = 0$. A more interesting case of zero correlation is when $y(t)$ is a noise waveform.

The function defined above is the *normalized* correlation coefficient, much used in statistics. In signal processing, it is frequently sufficient to work with an *un*normalized correlation coefficient, in which the denominator is omitted, and this will be so in what follows.

Evidently, to determine whether two signals are correlated, only two operations have to be performed – multiplication and integration (or addition plus storage of a running sum). We have already met a device of this type in the phase-sensitive detector, which finds the degree of correlation between the input signal waveform and a reference square wave. However, a correlator is a more general type of instrument, which accepts two arbitrary input signal waveforms and determines the degree of correlation between them. In addition, we now see that a matched filter is a type of correlator. In fact it correlates an incoming signal with a "hardwired" internal reference or idealized form of the signal it is required to recognize. It is important to note that such devices are expected to produce time-varying output waveforms showing degree of correlation as a function of time. Equation (15.2) is not a good model of this process which is typically performed by integration over a *restricted* period of time, i.e. the limits of the integrals are $t - T$ and t, where T is the averaging period (see Section 14.4 for a more general analysis of linear signal averaging circuits).

Thus the output of any correlator or matched filter is an output waveform which is inspected closely for peaks indicating high correlation between the compared waveforms. Either an observer or an electronic decision circuit will have to decide on the existence or significance of any peaks. A simple approach is to employ a local peak detector and a device for determining whether the peak is above a certain threshold.

Matched filters and correlators basically perform multiplication and integration operations, though (as mentioned above) integration amounts to an addition plus storage operation. Naturally, these arithmetic and storage operations can be carried out by analogue or digital circuits. It is worthwhile to think carefully in a given application which of these techniques might be more appropriate, since each has its own advantages: accuracy, duration of the signal, expense and compatibility with other equipment all have to be taken into account. These considerations are also relevant for signal averagers and lock-in amplifiers.

15.4 Special matched filters for radar

One of the main problems in radar is how to achieve optimal temporal resolution of the echo signal. Another is that of optimizing target detectability at long ranges. Both of these problems were tackled quite early on by the rather subtle technique of pulse compression. In fact this technique borrowed a similar method already used by bats for ultrasonic (acoustic) echo location: this involves sending out a "chirp" signal, which is a pulse with a frequency-modulated carrier. We can model the process quite simply as sending out a pulse with a frequency which increases linearly until the end of the pulse (Figure 15.6). When the echo returns, the first frequency to be received is the lowest, and thereafter the frequency increases until the end of the echo. If, however, we could delay the lower frequencies by appropriate amounts, all the power in the echo would arrive instantaneously, and the resulting SNR would be exceptionally large. Thus we would have used the varying frequency of the carrier as a parameter by which to concentrate the energy of the pulse and to improve the SNR.

This approach clearly amounts to coding the original signal to improve its detectability, together with use of an appropriately designed matched filter by which to achieve optimal detection. Note that the required filter does indeed fulfil the requirements of a matched filter, in that the impulse response is the time-reverse of the outgoing signal.

The chirp pulse compression method described above is highly successful in practice, but it has a major disadvantage – that it can give misleading

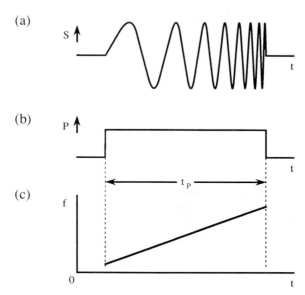

Fig. 15.6 Radar chirp signal. (a) A radar chirp signal, as transmitted (ideally, the received signal would be identical, but in practice noise would be present); (b) the transmitted power level; (c) the instantaneous frequency as a function of time.

results for moving targets. This is because of the Doppler effect. For example, with a target that is moving at a steady velocity v towards the receiver, the frequency is increased by the factor $(1 + 2v/c)$, where c is the velocity of light. This means that the received signal will be artificially delayed less than it should be by the chirp matched filter; thus the time-response will be misleading – even though the pulse will be compressed just as it should be. We can draw a graph showing the resulting response as a function of relative delay and Doppler shift for chirp signals of this type; the result is called the *ambiguity function* and is shown in Figure 15.7 for the above case. Notice that this function is not a point delta function at the origin but is a line through the origin showing a range of ambiguities between Doppler shift and delay. This graph should be compared with the ambiguity function for a normal non-frequency-modulated pulse (Figure 15.8).

It turns out that we have made a slight error in Figure 15.7, since we have assumed that all incoming frequencies will be accepted and shifted linearly by the matched filter. In fact, a true matched filter would only accept frequencies within the transmitted range of frequencies, and so those in the echo that are too high or too low will be ignored, and this will result in a triangular response pattern in the ambiguity function (Figure 15.9). Overall,

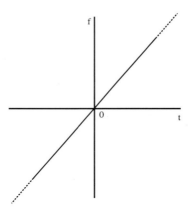

Fig. 15.7 Ambiguity function for a chirp signal with linear frequency variation. Here the graph shows the values of relative delay and Doppler shift for which a non-zero response is obtained from a filter matched to the chirp signal of Figure 15.6.

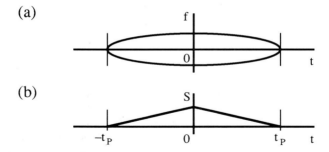

Fig. 15.8 Ambiguity function for a pulse without frequency modulation. Here (a) shows the ambiguity function for a conventional non-frequency-modulated pulse. In fact (a) shows only the approximate *extent* of the pulse in the time and frequency domains, while (b) shows the cross-section of the two-dimensional function along the $f = 0$ ("zero Doppler") axis. Notice that the response in (b) is the autocorrelation function of a square pulse of duration t_P, which results from assuming a matched filter response to the square pulse signal. The non-zero extent of the signal along the $t = 0$ axis in (a) arises since the frequency of a carrier wave is not well defined if the wave lasts only a time t_P (see text).

the effect of pulse compression is to *shear* the ambiguity function along the frequency axis (i.e. to give a shift along the frequency axis that is proportional to the distance from the origin measured along the time axis), and this has the effect of very significantly reducing its width along the time-delay axis –

(a)

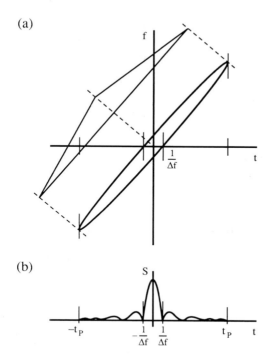

(b)

Fig. 15.9   Detail of the ambiguity function for a chirp signal. This figure shows more detail of the ambiguity function of Figure 15.7. First, the function does not have infinite extent since no part of the signal can be moved more than $t_P/2$: in fact, the overall extent in the time domain is the same ($-t_P$ to $+t_P$) as for a non-chirp signal. However, the ambiguity function is best imagined as a form of the original ambiguity function that has been sheared parallel to the frequency axis. Hence the central part of its cross-section along the time axis is significantly narrowed (b).

though, as indicated above, for this mechanism to work perfectly there must be no Doppler shift in frequency.

It can be shown that the width of the ambiguity function along the time-delay axis (measured between the most central pair of zero-crossings) is approximately $2/\Delta f$, whereas the width of the ambiguity function for the normal non-chirp pulse is $2t_P$, where t_P is the pulse width. Hence the *compression ratio* is:

$$D = t_P \Delta f \tag{15.3}$$

It is fortunate that the final width of the echo bears no relation to its intrinsic width $2t_P$, but merely depends on how wide we can make the frequency modulation on the transmitted pulse. Typical values of D are in the range 50–500, pulses being compressed from $\sim 20\,\mu s$ down to much less than $1\,\mu s$ by devices such as SAW filters (see Appendix B).

An interesting aspect of this approach is that we have only been able to distort the ambiguity function, so that its zero-frequency cut has a narrow profile. As a result, the function has been extended to much higher frequency shifts Δf. This is a consequence of the fact that the total volume underneath the (2-D) ambiguity function must be constant. However, the particular way in which the reshaping of the ambiguity function has been achieved in this case will frequently be disadvantageous: in general a better strategy is to aim for a "thumb-tack" response, in which the unwanted volume in the ambiguity function is spread as uniformly as possible over the whole parameter space (in this case the frequency-time parameter space). Then there will be a minimized possibility that misleading signals could arise from any moving targets. Numerous ways have been invented for attempting to achieve such an ambiguity function, and none of them is perfect. In the next sub-section we consider the Costas frequency coding approach.

15.4.1   Costas frequency coding

In the Costas frequency coding method, it is recognized that linear coding of frequency with time is the poorest possible choice. To understand the reasons for this, we proceed to a *frequency hopping* situation in which we jump from one frequency to another in steps, throughout the period of the transmitted pulse. Clearly, linearly increasing frequency coding is only one of $n!$ codes in which frequencies can be applied in the n available time-slots. There is no reason to believe that linear frequency coding is especially effective. A better scheme might be to take a random distribution of frequencies during the period of the transmitted pulse. We can get a reasonable approximation to this by ensuring (a) that there is just one frequency per time-slot, (b) that all of these frequencies are different, and (c) that the vector $(\delta f, \delta t)$ is different for all pairs of frequencies.

We illustrate these concepts by a simple example. First, take the case shown in Figure 15.10(a) corresponding to a linear frequency coding. This is displayed as a 3×3 matrix with one non-zero element per row and one non-zero element per column. This clearly fulfils condition (a) above; condition (b) is also fulfilled since each non-zero element in the matrix corresponds to a different frequency. However, condition (c) is not fulfilled,

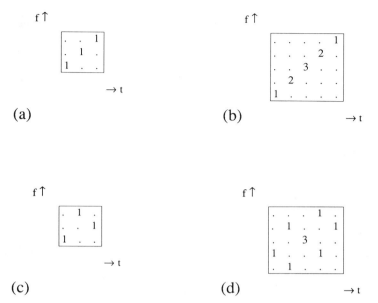

Fig. 15.10 Costas frequency coding. This figure shows various frequency-hopping signals (signals for which the frequency changes in steps between one time-slot and another) and their responses. (a) A case of linear frequency coding; (b) the resulting response; (c) a case in which the frequency coding is non-linear. Note that the response (d) for (c) is much closer to the ideal "thumb-tack" response than the highly structured response shown in (b).

and as a result we find that the ambiguity function takes the highly structured form shown in Figure 15.10(b) (notice the similarity of this figure to Figure 15.7 – see also Figure 15.9). If we now re-arrange the order of the frequencies to obey condition (c) (see Figure 15.10(c)), we obtain the ambiguity function shown in Figure 15.10(d).

The interesting thing about the final ambiguity function is that there is one central peak, whose size is identical to that for the linear coding case, but there are no other peaks of size greater than unity. In addition, the sidelobe peaks are distributed fairly evenly over the f–t parameter space and the response pattern, though symmetric about the origin, has little structure (it is as close to "random" as it could be for this region of the parameter space). These observations apply in other cases of Costas coding – see for example the case of five hopping frequencies shown in Figure 15.11. In general, we can quantify the number of peaks and their sizes and positions

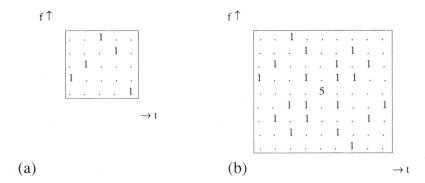

Fig. 15.11   A more complex case of Costas coding. This figure shows a more complex case of Costas coding (a) and its response (b).

by noticing the following:

1. The size of the pulse parameter space is $n \times n$.
2. The size of the signal parameter space is $(2n-1) \times (2n-1)$.
3. The number of responses cast in the signal parameter space is n^2.
4. The number of these responses that contribute to the main peak is n.
5. The size of each sidelobe peak is 1.
6. The number of sidelobe peaks is $n^2 - n$.
7. The total number of peaks is $n^2 - n + 1$.
8. The total number of null locations is $(2n-1)^2 - (n^2 - n + 1) = 3(n^2 - n)$, which is equal to 3 times the number of sidelobe peaks.
9. The number of responses per row, and per column, obeys a triangular variation: $1, 2, 3, \ldots, n, \ldots, 3, 2, 1$.

Overall, we see that the result is quite close to the required thumb-tack response, though the fact that there are so many null locations spreads out the ambiguity function over a larger parameter space than might appear desirable. (On the other hand, any attempt to compress the response further is quite likely to give larger sidelobe peaks, so there is bound to be a limit on how compact an ambiguity function is achievable.) It turns out that if the (ordered) frequencies in the n time-slots have values $f_i = i\delta f$ $(i = 1, \ldots, n)$, then the overall spread of time-delays in the ambiguity function is $\pm n/\delta f$, while the overall spread of frequencies is $\pm n\delta f$. Clearly, while the proper way to lower the pedestal of the thumb-tack response is to increase n, this also has the effect of increasing the overall spread in time and frequency space: this is because the volume of the ambiguity function is constant and any displaced volume has to go somewhere. Such arguments indicate that there are fundamental physical limitations on how good the response can be.

15.4.2 Barker codes

An alternative approach to the coding of pulses in order to achieve high resolution echoes is via phase coding. This is in some ways similar to frequency coding, but it is rather more easily achieved in many radar systems, since it is only necessary to change the phase of the outgoing radiation by switching in suitable time-delay lines.

Barker codes employ rather simple binary phase-coding sequences, in which the phase is inverted for certain time-slots (Figure 15.12). Again, the idea is to employ a sequence for which the amplitudes are constant, and which give a final signal with sidelobe peaks that are no larger than $1/n$ of the main peak. Perhaps oddly, only nine such sequences are known. They are shown in Table 15.1.

In each of these cases the response (autocorrelation function) is also given. Clearly, these codes obey the stated requirements. However, their performance characteristics under Doppler shifts from moving targets are not clear

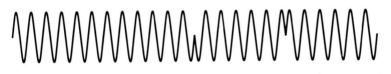

$\longrightarrow$
t

Fig. 15.12 A Barker coded waveform. This figure shows a typical Barker coded waveform, corresponding to the case $+ + - +$ for which $n=4$ (see Table 15.1): for every "$-$" element in the code, the phase is inverted.

Table 15.1 Barker phase sequences and their response functions.

n	Barker sequence	Response
2	$+ +$	1 2 1
2	$+ -$	-1 2 -1
3	$+ + -$	-1 0 3 0 -1
4	$+ + + -$	-1 0 1 4 1 0 -1
4	$+ + - +$	1 0 -1 4 -1 0 1
5	$+ + + - +$	1 0 1 0 5 0 1 0 1
7	$+ + + - - + -$	-1 0 -1 0 -1 0 7 ...
11	$+ + + - - - + - - + -$	-1 0 -1 0 -1 0 -1 0 -1 0 11 ...
13	$+ + + + + - - + + - + - +$	1 0 1 0 1 0 1 0 1 0 1 0 13 ...

from the above ideas. It has been shown elsewhere that these codes do not give such high detectabilities for Doppler-shifted signals, though in addition, they do not yield interfering signals for high frequency shifts, and are therefore less likely to give misleading responses. Thus it seems that Barker codes are reasonably well adapted to the detection of non-moving targets, and are somewhat less liable to be muddled by the presence of moving targets.

15.5 First-principles calculation of an optimal filter

In this section we compute the type of weighting template that is required to optimally enhance* a signal that is subject to noise. We assume: (1) a set of discrete times t_i at which signal levels are definable; (2) signal levels S_i that are zero except in the vicinity of a pulse to be detected; (3) an n-element linear weighting template with coefficients w_i for enhancing signals; (4) noise levels that are subject to independent local distributions with standard deviations N_i.

The total signal received from the weighting template will be:

$$S = \sum_{i=1}^{n} w_i S_i \tag{15.4}$$

while the total noise received from the weighting template will be characterized by its variance:

$$N^2 = \sum_{i=1}^{n} w_i^2 N_i^2 \tag{15.5}$$

Here we have added variances from the individual noise sources as they are assumed totally uncorrelated. Hence the (power) SNR is:

$$\rho^2 = S^2/N^2 = \left(\sum_{i=1}^{n} w_i S_i\right)^2 \bigg/ \sum_{i=1}^{n} w_i^2 N_i^2 \tag{15.6}$$

For optimum SNR, we need to set $d\rho^2/dw_i = 0$ for all i. First we compute

---

* Technically, an optimal filter of this type only *enhances* a signal, which may then be detected by thresholding and/or peak location.

the derivative

$$d\rho^2/dw_i = (1/N^4)[N^2(2SS_i) - S^2(2w_iN_i^2)]$$
$$= (2S/N^4)[N^2S_i - S(w_iN_i^2)] \tag{15.7}$$

Next we require that $d\rho^2/dw_i = 0$. Hence

$$N^2S_i = S(w_iN_i^2) \tag{15.8}$$

Since this is valid for all i, we may average over all n elements in the weighting template:

$$N^2\overline{S} = S(\overline{w_kN_k^2}) \tag{15.9}$$

Hence

$$\frac{S_i}{\overline{S}} = \frac{w_iN_i^2}{\overline{w_kN_k^2}} \tag{15.10}$$

$$\therefore \quad w_iN_i^2 = \left(\frac{S_i}{\overline{S}}\right)\overline{w_kN_k^2} \tag{15.11}$$

$$\therefore \quad w_i = \left(\frac{S_i}{\overline{S}}\right)\frac{\overline{w_kN_k^2}}{N_i^2} \tag{15.12}$$

In fact it is more revealing to write this equation in the form:

$$w_i \propto \frac{S_i}{N_i^2} \tag{15.13}$$

It is of interest to try to understand this equation in detail, and in particular to explain why there is an inverse *square* dependence on noise. Let us first suppose we have a case in which all the N_i are equal and have value N_c. Then we have the simple relation:

$$w_i \propto S_i \tag{15.14}$$

which is intuitively correct.

Next take a case in which the N_i are not equal. Let us take steps to make them equal by multiplying the individual samples by an appropriate factor – which must clearly be proportional to $1/N_i$. Then the resulting noise level will be multiplied by the factor N_c/N_i, and the signal samples will also be multiplied by the factor N_c/N_i. Thus we obtain the new signal levels:

$$S_i' = S_i \times N_c/N_i \qquad (15.15)$$

We can now apply the simplified formula (equation 15.14), so the weighting factors will be:

$$w_i' \propto S_i' \qquad (15.16)$$

and the *overall* weighting factors will be:

$$w_i = (N_c/N_i) \times S_i' = S_i \times (N_c/N_i)^2 \qquad (15.17)$$

This confirms our earlier general formula, equation (15.13). In addition, it shows that we can consider optimization of the SNR as the application of a noise-equalizing filter (which affects both the noise and the signal) and a simplified form of optimal filter.

In fact, there are now three cases which it is worth considering. The first is that of white noise, and in that case the noise at the individual sampling points is uncorrelated and can normally be taken to be independent and subject to identical amplitude distributions. In that case all the N_i in the above theory are equal and equation (15.14) applies. The second case is that of non-white (e.g. $1/f$) noise, and is not covered immediately by the above theory: it will be considered separately below. The third case is where the noise has a known temporal distribution and *is* covered by the above theory, the N_i having different but known values. Although this last case does not arise very frequently, it is worth considering – e.g. when a radar echo (Chapter 16) or magnetic spin-echo (Chapter 17) is initiated by a strong microwave pulse which affects the receiver and gives rise to a decaying noise component: this may be eliminated by crude temporal switching, as in a boxcar detector, but in theory at least, carefully chosen temporal weighting factors should be used to retain as much of the signal as possible (see equation 15.13).

15.5.1 Optimal filters for the case of non-white noise

In the main part of Section 15.5, we obtained a formula for the weights of an optimal filter that was valid in the case of white noise. Such a filter is called a *matched filter*. Here we consider how to construct an optimal filter when the noise is not white. In fact the process is quite simple: all we need to do is to make the noise white by applying a *noise-whitening filter*, noting that this will also affect the signal; then we detect the transformed signal using a matched filter (this can now validly be applied since the noise is white).

At this point it is easier to pursue the problem mathematically in the continuous case where the signals are not sampled. Let the signal have time development $s(t)$ and Fourier transform $S(f)$, and suppose the noise has power spectrum $|N(f)|^2$. Applying a noise-whitening filter $W(f) = w/|N(f)|$ produces frequency domain signal and noise voltages respectively of $S'(f) = W(f)S(f) = wS(f)/|N(f)|$ and $|N'(f)| = W(f)|N(f)| = w$, where w is a constant. We now deduce that the matched filter of the noise-whitened signal has the form $S'(f)$, with the result that the *overall* filter for optimizing the SNR has the form:

$$S''(f) = W(f)S'(f) = w^2 S(f)/|N(f)|^2 \qquad (15.18)$$

Note the similarity of this equation to equation (15.13): whereas the latter is an equation of temporal signals and noise and is necessary when dealing with noise whose amplitude distribution has a known variation in the *time* domain, equation (15.18) relates to signals and noise whose amplitude distributions have known variations in the *frequency* domain. Thus there is an interesting symmetry between the two cases.

15.6 Summary

This chapter has studied the concept, design and application of matched filters. After trying various simple weighted time-delay filters, it showed that filters having an impulse response equal to the time-reverse of the signals to be detected are the most effective at enhancing SNR: such filters are called matched filters. Correlation was also considered, and it was found that matched filtering is a special case of correlation wherein the signals are correlated with a hardwired internal version of the impulse response. Next, special matched filters for radar were considered, including the linear frequency-modulated "chirp" pulse receiver, the Costas coded frequency-hopping technique (which approximates the desired thumb-tack response much more closely), and the Barker coded phase-hopping technique – all of which were found to be capable of increasing the output SNR by significant factors, while maintaining the requisite time resolution. (In fact, use of chirp and the other techniques described here can be used (a) to increase SNR, (b) to improve temporal resolution, or (c) to reduce peak transmitter power – or some combination of these.) Finally, a proof of the (optimal) form of the matched filter and associated noise-whitening filter was presented.

Whilst the special case of radar was dealt with in fair detail, matched filters for other applications were not considered. However, Chapter 17 provides case studies covering the topic of magnetic resonance, while Chapter 19 shows that the p.s.d. and boxcar detector are forms of matched filter, and discusses the consequences of this for optimal detection.

15.7 Bibliography

As a general rule, electronics books do not cover matched filtering, and the topic is left to volumes on signal and communication theory, such as those by Rosie (1966) and Schwartz (1980), or those on radar, such as Skolnik (1980) and Levanon (1988). Rosie provides further information on matched filtering, starting with the topic of correlation, while Wilmshurst (1985) provides theory, but with few supporting examples. Levanon significantly expands the study of matched filtering in radar, and uses the Schwarz inequality to prove the form of a matched filter in the case of continuous signals, but otherwise ignores background theory and the need for noise-whitening filters. However, Wilmshurst covers the latter topic quite thoroughly and Whalen (1971) gives a full theoretical treatment. Readers may be interested to learn that the whole subject of matched filtering was developed by North (1943) and others as late as the Second World War, in the context of radar – see also the review article by Turin (1960) for a summary of this early work. For the original paper on chirp radars, see Klauder et al. (1960), and for a popular treatment, see Wyndham (1968). For the original papers on the frequency- and phase-hopping techniques described in Section 15.4, see Barker (1953) and Costas (1984).

15.8 Problems

1. For the situation in Problem 14.1, derive the matched filter for the expected echo signal, and deduce the shape of the signal after it has passed through the filter. Confirm that the matched filter gives a better SNR than the optimized linear gate (Problem 14.1).
2. When a certain aircraft is flying directly towards a radar antenna, the echo has the profile shown in Figure 15.13. Give (a) the general design of a matched filter for this case, (b) its impulse response and (c) the time-response of the matched filter to the incoming signal. Compare the SNR achievable (a) when using the matched filter, (b) when the signal is

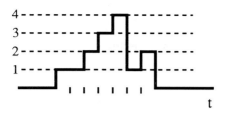

Fig. 15.13

merely averaged over the duration of the echo, and (c) when the signal is averaged over some optimal period.

3. Obtain a Costas signal for the case $n = 7$. Determine the ambiguity function that will result in this case. Check statements 1–9 made in Section 15.4.1 about Costas signals for this particular code. Show that for a Costas signal, the pulse compression ratio is equal to n in the general case.

4. Consider what form the signal takes when a Doppler shift occurs in Costas coding. Does the signal disappear? Is the Doppler shift ignored? Is there a largest allowable Doppler shift? Is the Doppler shift information measurable at the output of the decoder?

16

Radar

16.1 Introduction

Radar is an acronym for *radio detection and ranging*. It is a technique that was developed during the Second World War under the pressure for detecting, locating and ranging enemy aircraft, in the hope that they would be intercepted before they could do any damage. In fact, the concept of radar had been invented before the war, but it was during the war years that the necessary high-power microwave transmitters such as magnetrons* were produced and reliable crystal rectifiers were developed for signal detection.

16.2 Basic concepts

The basic concept of radar is to transmit a strong electromagnetic wave that will be reflected from a target such as an aircraft and which can then be detected by electronic receiving stations (Figure 16.1). As we shall see below, the echo signals received from aircraft are especially difficult to detect, as they involve *pico*watts of power, even though the microwave pulses generated by the magnetron valve in the transmitter typically involve many *mega*watts of power. However, the pulses can be made quite short (around 1 μs), as this helps the temporal and spatial resolution of the radar. (In fact, normal radar pulse lengths are in the range 0.1–5 μs, though in pulse compression radar systems such as those outlined in Section 15.4, the initial pulses may be as long as 50 μs.)

---

* For relevant details of microwave devices, see Appendix C.

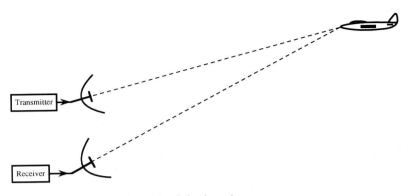

Fig. 16.1 A basic radar system.

We now consider the range of signal power levels that arrive at the receiver. Radar waves, like other forms of electromagnetic radiation, are inherently subject to the inverse square law of attenuation. Thus only a small fraction of the transmitter power arrives at any target, and of this radiation, only a fraction of the reflected power travels back towards the receiver – i.e. the inverse square law operates twice. If the receiving antenna (aerial) is adjacent to the transmitter antenna, or the same antenna is used for both purposes, then overall there will be an inverse *fourth* power law of attenuation, i.e. the echo signal will have been attenuated by a factor $1/R^4$, where R is the distance to the target. Now range is one of the prime considerations with radar systems. It is always desirable to press the range of a radar system as far as possible. Yet we see from the $1/R^4$ distance attenuation factor that doubling the range of a radar system requires ~ 16 times the transmitted power. Hence, great care has to be taken in the design of radar systems to cut out power losses and to design sensitive receivers capable of detecting signals well down into ambient noise levels. Unfortunately, it is not just noise that a radar system has to contend with, but also clutter signals from irrelevant objects, such as buildings near to the transmitter. Much attention therefore has to be paid to cutting down clutter effects. In addition, considerable care has to be taken to prevent the receiver from being saturated by the original high-power transmitted pulses. Fortunately, such effects can be reduced by careful attention to the relative timings of the transmitted and received signals.

16.3 Implementation of a radar system

We now consider in more detail the design of a radar system. Inevitably, it will be necessary to save on cost, and in practice this means that the same antenna will normally be used for transmission and reception. This makes it imperative to employ some effective means for preventing the transmitted pulse from saturating the receiver. This is achieved by means of a TR-cell, which is a special flash tube that is placed in the waveguide in the path of the received signal: when the high-power transmitted pulse arrives, this causes the TR-cell to conduct and to short out the pulse so that it is highly attenuated before it can proceed to the receiver (Figure 16.2). In fact, *some* breakthrough may get to the receiver, especially in the time before the TR-cell is fully conducting. The effect of this can be minimized (a) by using some means for pre-igniting the TR-cell, (b) using PIN diodes in the waveguide after the TR-cell, and (c) by switching the receiver in some way. However, we do not go into further details here.

Next, there is the problem that the intrinsic sensitivity varies widely for different ranges, because of the variation in the $1/R^4$ attenuation factor. This means that the gain of the amplifier has to be "shaped" (Figure 16.3) as a function of time delay T after the radar pulse is transmitted – this is, of course, possible because T is proportional to R. While this technique can be made to work satisfactorily, it cannot compensate fully for the effects of scatter from buildings, trees or other objects near to the transmitter. In the end, clutter signals must be eliminated electronically (or by computer) on examining the signals emerging from the receiver. Fortunately, in many cases there will be a particular target of interest, and this will not vary much from one radar scan to the next. Hence it will be possible to filter clutter signals electronically using a boxcar detector arrangement, simply by switching a linear gate on and off at appropriate instants (Figure 16.3).

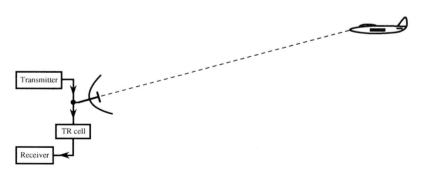

Fig. 16.2   Use of a TR-cell to eliminate direct radiation from the transmitter. This figure shows a TR-cell being used to eliminate direct radiation from the transmitter to the receiver. This is especially important when the same antenna is used for both purposes.

(a)

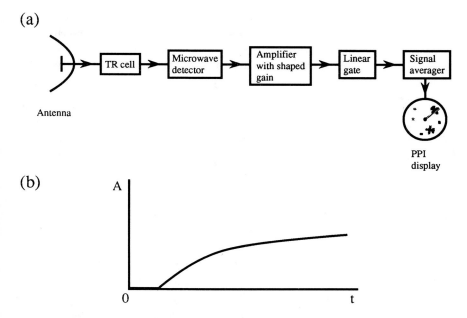

(b)

Fig. 16.3 Block diagram of radar receiver. In (a), an amplifier with shaped gain A (b) is used to allow for the $1/R^4$ attenuation factor, and to minimize the effects of clutter. The final signal is shown on a standard PPI (plan-position indicator) display.

We now get to the important problem of how to optimize sensitivity. The first means of improving sensitivity is to perform signal averaging, in particular by averaging a sequence of echo signals. With a pulse radar system, this is easily achieved by transmitting high-power microwave pulses at a steady rate of perhaps 1000 per second. In fact pulse repetition rates of 1 kHz are suitable for locating targets up to ~ 100 km away, since radar waves travel with the velocity of light, some 3×10^8 m/s (i.e. 3×10^5 km/s, or 300 m/μs). Notice that an aeroplane travelling at ~ 700 km/hour travels some 100 m in 500 ms – so if its path is to be plotted on a radar screen at a resolution ~ 100 m, only ~ 500 echoes can be averaged for each distinct position.

A further means of increasing sensitivity is to make the receiving system sensitive to echo pulses of the expected shape and duration. Let us model an "average" echo signal as being of similar shape to a Gaussian distribution function, but with fairly sharp cut-off near the $\pm 3\sigma$ points (the *original* echo cannot start before the radar wave strikes the leading edge of the target, and cannot terminate after the wave has gone past the target). Then we can use a linear gate to cut out noise and clutter outside this range of time delays*. However, at the extremes of this range the signal will be quite low, and the noise will be at its normal background level. Hence we can beneficially eliminate a little more of the echo signal, in the knowledge that we are eliminating a greater proportion of noise than signal (Figure 16.4). In fact, there is an optimum level of cutoff, and for white noise this is readily calculable (see Section 14.6). However, we can do much better than this if we employ a matched filter detection system. In that case, the small parts of the echo which we were previously prepared to lose can usefully be retained but need not add substantially to the total noise, since a suitable weighting factor is included. As we have already seen, the matched filter is designed specifically for situations where white noise is present, and this means that every sample of the input signal has the same local noise probability distribution with the same standard deviation σ (Figure 16.4) – a condition that does not apply for clutter but is a good approximation for most true noise inputs. Hence in many cases use of a matched filter should come close to optimizing the final SNR. Although an overt matched filter should in principle always be used, it may sometimes be more convenient to approximate the matching condition by giving the i.f. (intermediate frequency†) amplifier of the radar detection system a shaped passband whose frequency response function is the Fourier transform of the desired time-domain response.

16.4 Summary

This chapter has given a brief resumé of radar technology, showing in particular that the $1/R^4$ law gives the echo signals an exceptional dynamic range, while leaving the receiver with a rather disastrous SNR problem. Thus superheterodyne reception, matched filtering, signal averaging and clutter

---

* However, the purpose of radar is generally to *search* for echoes, so in practice we have to sweep the linear gate over all times, or use a series of linear gates, or alternatively employ a moving-average filter of the same time duration. Such a filter can be implemented as a transversal filter.

† In radio and radar it is common to convert incoming high-frequency signals at or around the carrier frequency down to some intermediate frequency which is much easier to handle and to amplify before final signal detection takes place – this technique being commonly known as superheterodyne detection. In fact, this approach allows much more accurate tuning of the carrier frequency, and much improved shaping of the passband: it also ensures that signal detection takes place under near-ideal conditions, where the diodes have substantial signals to handle. In radar, the i.f. is typically ~ 30 MHz (compare ~ 465 kHz for a.m. radio and 10.7 MHz for v.h.f. radio) – see Chapter 18.

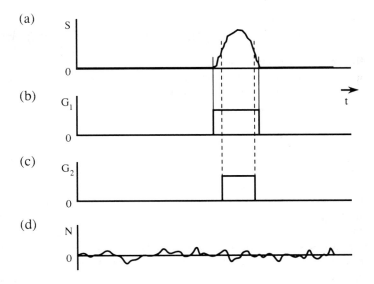

Fig. 16.4 Optimal gate width for a radar echo. (a) A typical radar echo; (b) the obvious gate width; (c) a gate width that will improve the SNR (note that a matched filter is required to *fully* optimize the SNR); (d) a typical noise waveform.

rejection are all vital considerations in this subject. The chapter should be studied in conjunction with Appendix C, which covers some relevant aspects of microwave technology. However, the treatment of microwaves, and of transmitter technology, is necessarily curtailed for lack of space, and the chapter is targeted particularly at signal recovery aspects of radar. Note also that some of the detail of matched filtering for radar is covered in the later sections of Chapter 15.

16.5 Bibliography

For a comprehensive study of radar, the reader is referred to the classic book by Skolnik (1980), but a recent short readable (non-theoretical) account is presented by Lynn (1987). Levanon (1988) provides a more thoroughgoing study of various theoretical aspects of radar, and in particular covers chirp and other radars in some depth, including much detail on the design of relevant forms of matched filter. However, the latter volume largely ignores practical aspects of the subject. In fact, practical aspects of radar devolve to a large extent into details of microwave systems. For the reader requiring more detail than is presented in Appendix C, the short readable study by Matthews and Stephenson (1968) should be useful, whereas for a fuller, more up-to-date and theoretical treatment, Chatterjee (1988) may be consulted. It is also worth mentioning the various exhilarating historical accounts of the wartime development of radar by those who took part in it – for example the famous physicist Hanbury Brown (1991).

16.6   Problems

1. Estimate the amount of power in a radar echo, if a 1 MW pulse is radiated by the transmitter, and an aeroplane of visible cross-section $15\,\text{m}^2$ is to be detected at a distance of $10\,\text{km}$. Assume that the transmitting antenna has a beam-width of $2°$, the aircraft reflects radiation uniformly over an angle of $100°$, and that the receiving antenna has an area of $0.3\,\text{m}^2$.

2. The *gain* of a radar antenna is defined as the signal (power) improvement factor resulting from the directivity of the antenna. If the horizontal and vertical beam-widths of the antenna are $\delta\theta$ and $\delta\varphi$ degrees respectively, determine the ideal gain of the antenna. Calculate the value in dB when $\delta\theta \approx 2°$ and $\delta\varphi \approx 12°$. (Note that the efficiency of a radar antenna is normally only $\sim 50\%$, thereby reducing the gain by a factor of about 2 relative to the ideal.)

3. Estimate the minimum signal detectable by a radar receiver, starting with the expression $kTBF$ for the effective input noise power level, where $k = 1.38 \times 10^{-23}\,\text{J/K}$, $T = 290\,\text{K}$, $B = 10\,\text{MHz}$, and $F = 2\,\text{dB}$. Refine your estimate to take account of at least 25 "blips" being averaged visually on the radar screen, and to ensure that there is a SNR of $15\,\text{dB}$ to limit the false alarm rate. Show that your result is consistent with a rule-of-thumb figure for the minimum detectable radar signal of $10^{-13}\,\text{W}$.

17

Magnetic Spin-echo Systems

17.1 Introduction

This chapter is about the phenomenon of magnetic spin-echoes and how they are detected. It will be directly relevant to many physicists, chemists and biologists, who may well need to study the topic in some depth. However, engineers may not be so concerned with the specific concepts underlying spin-echoes*, though they will probably be particularly concerned with the lessons on how sensitivity is optimized for such processes. Readers who are in this position can therefore take the present chapter as providing useful case studies of the process of signal recovery: in that case much of Section 17.2 and some of the details of later sections can be skipped, provided that the results of Figure 17.4 are borne in mind at a phenomenological level.

17.2 The basic concept

There are many types of impurity that can appear in crystals such as aluminium oxide or calcium fluoride and which have interesting magnetic properties. These properties arise because when the impurity ion (e.g. Fe^{3+} or Cr^{3+}) is substituted in the crystal lattice for the host metal ion (e.g. Al^{3+} or Ca^{2+}), it has one or more electrons to spare. These electrons are bound to the local impurity ion and orbit around it imparting the well-known magnetic effect of a current loop. In addition, the quantum mechanical "spin"

---

* Unless they are involved in the manufacture of Magnetic Resonance Imaging body scanning equipment!

241

of the electron has a magnetic effect which complements the "orbital" magnetism. Overall, when considering crystals containing these magnetic impurities, we can imagine that there is a single resultant magnetic spin at each site. In what follows we consider the effects of the individual magnetic spins, and ignore the precise physical processes that give rise to them.

It is well known that a free magnetic moment $\mathbf{M}$ will precess about a magnetic field $\mathbf{H_0}$ in which it is placed (Figure 17.1). In many cases fields of about $10\,\mathrm{kG}$ ($1\,\mathrm{T}$) cause the magnetic spins to precess about $\mathbf{H_0}$ at some $10\,\mathrm{GHz}$ ($10^{10}\,\mathrm{Hz}$), which is in the microwave $3\,\mathrm{cm}$ band (commonly called *X-band*). Thus electromagnetic waves of precisely the right frequency will cause the magnetic spins to gain energy – or in quantum mechanical terms, will cause them to "flip" (see also Figure 13.11).

It is easy to visualize what happens if we jump into a frame of reference that is rotating about $\mathbf{H_0}$ at the precession frequency of the spins. In this frame, $\mathbf{H_0}$ can be ignored since the precession it causes has already been taken into account. In addition, the electromagnetic wave field is now a constant $\mathbf{H_1}$ and points at right angles to the direction of $\mathbf{H_0}$ (Figure 17.2). In the rotating frame the spins therefore see only the field $\mathbf{H_1}$ and precess around this new direction. Hence they no longer remain aligned along the direction of $\mathbf{H_0}$ but rotate away from it (Figure 17.2). This means that, in the *laboratory* frame of reference, they appear like small rotating magnets (as in Figure 17.1) and thus radiate microwave signals that can be picked up by a sensitive receiver.

It is interesting that this effect is largest when the spins are rotated by exactly 90° from the "spin up" direction so that they give the greatest possible rotational magnetic field. A microwave $\mathbf{H_1}$ pulse that rotates the spins through exactly 90° is called a $\frac{\pi}{2}$-pulse. We next consider what happens after a $\frac{\pi}{2}$-pulse. At this point two things happen. First, the spins start dephasing in the

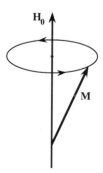

Fig. 17.1   Free magnetic moment $\mathbf{M}$ precessing in a magnetic field $\mathbf{H_0}$.

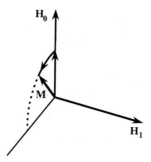

Fig. 17.2 Magnetic spins precessing about the oscillating field H_1. This figure depicts what happens in a frame of reference that is rotating about the direction of the constant magnetic field H_0 with angular velocity equal to that of the oscillating field H_1. In this frame H_0 can be ignored and H_1 appears to be stationary. Thus **M** rotates about the resultant field direction H_1.

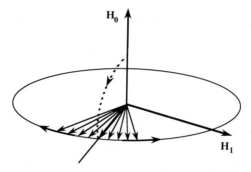

Fig. 17.3 Dephasing of spins after a $\frac{\pi}{2}$-pulse. Here, the magnetic spins have been rotated through an angle of 90° in the rotating frame, as indicated in Figure 17.2; they then start dephasing in the equatorial plane, some moving forwards and some backwards, under the influence of slightly differing local H_0 fields – see the contra-rotating arrows.

rotating frame, some moving forwards and some moving backwards (Figure 17.3), until eventually they have fanned out uniformly and have no resultant magnetic component in the rotating frame. At this point they have no rotating component in the laboratory frame either, and they stop radiating microwave energy. However, although the spins have dephased, with relaxation time T_2, individually they remain pointing at 90° to the main magnetic field direction H_0, in the rotating frame. This situation does not apply for ever, and the spins gradually move back to the low-energy direction aligned along the main magnetic field direction $\mathbf{H_0}$, the total magnetic moment along $\mathbf{H_0}$ at time t being:

$$M = M_0[1 - \exp(-t/T_1)] \tag{17.1}$$

This process happens with a relaxation time T_1 which is generally much longer than T_2.

One of the remarkable things about the different precessions that happen after a $\frac{\pi}{2}$-pulse is that their effects are largely reversible. To achieve this we only need to institute a π-pulse, which has the effect of inverting the precessions that have already taken place; thereafter the spins continue to precess in the *same* direction as before, and the individual precession angles gradually unwind and go to zero – though after this they continue to dephase as they did originally*. Thus we get the effect shown in Figure 17.4, in which the original decaying signal from the spins‡ is reconstituted and produces a "spin-echo" after a predictable time. The π-pulse has clearly acted as a mirror which enables the original signal to appear again. In fact, the effect is not completely reversible, and the decay may be characterized by a decay time $T_2^\dagger$ which is reversible and another decay-time T_2 which is not reversible (Figure 17.4(c)). In any case the T_1 decay rate is irreversible. (It is quite easy to understand the physical meanings of these decay times: basically, after any disturbance, the spins first come to thermal equilibrium with each other, with *spin-spin* relaxation time T_2: then they return to thermal equilibrium with the crystal lattice, with *spin-lattice* relaxation time T_1. On the other hand $T_2^\dagger$ is due to a totally different effect – so-called *inhomogeneous broadening*, i.e. slightly differing H_0 magnetic fields appearing at each spin site: as the resonant frequencies of the spins are spread out in this way, the spins fan out slightly differently in the rotating frame. We do not need to give more attention to these details in the present context, since we shall concentrate below on the ease with which spin-echoes can be detected by electronic systems.)

Before leaving this section, it is worth remarking on what the reader may already have deduced – that it is possible to obtain many spin-echoes within one T_2 decay, by applying a sequence of π-pulses after the initial $\frac{\pi}{2}$-pulse: this scheme is a highly useful one (e.g. see Section 17.6), but further detailed discussion is beyond the scope of the present volume.

---

* Here there is an interesting analogy with runners in a race. At the gun the competitors start running, and at a certain point a second gun is fired and the competitors turn round and run at the same speeds in the reverse direction. If none of them tires they will all pass the starting line at the same moment!

‡ This signal is commonly known as a *free induction decay* (FID).

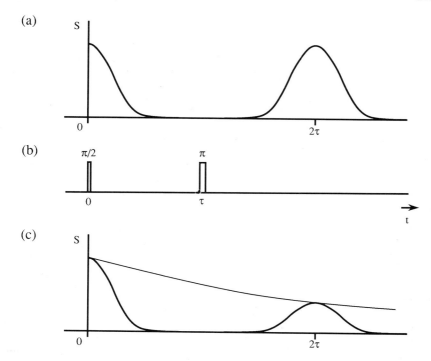

Fig. 17.4 Formation of a spin-echo. Here (a) shows first the decay of the signal from the rotating spins as they dephase in the rotating frame; it also shows how the phases become coherent again to form an echo signal after a π-pulse (b) has been applied. (c) In practice the spin-echo is reduced in size as a result of the time delay 2τ that has occurred since the original $\frac{\pi}{2}$-pulse.

17.3 Setting up an electron spin-echo system

In practical situations we require to optimize the signal that is obtained from the magnetic spins in a crystal, in order to study their environment in a crystal lattice. Whereas the latter is a problem in physics, we concentrate here on how the signal is detected by electronic detection apparatus. First, we need to produce microwave pulses of the correct duration to act as $\frac{\pi}{2}$- and π-pulses. Then we need to try to detect the spin-echoes. The latter are weak signals and are liable to be swamped by the lingering effects of the high-power microwave pulses – in a manner very similar to the case of radar echoes from aircraft. There are several ways in which these lingering effects can manifest themselves:

1. The microwave pulses themselves may not decay away quickly enough.
2. The electronic signals that produce the pulses may not have disappeared.
3. The detector crystals may not have recovered.
4. The amplifier following the detector may not have recovered.
5. The power supplies feeding the amplifier may not have recovered.

   The first of these problems is generally solvable by use of PIN diode switches in the microwave system. The second should not be a problem, given careful system design. The third is also not normally a problem. However, the fourth and fifth items taken together are the major source of difficulties that limit the effectiveness with which spin-echoes can be detected after short $\frac{\pi}{2}$- to π-pulse delay times τ. Overall, it is very helpful to employ a linear gate to help suppress the remaining effects of the microwave pulses (Figure 17.5): of course, it will also help suppress electronic noise that is picked up by the detector and amplifier. In addition, the linear gate can be included in a boxcar detection system, with the advantage that averaging of the spin-echoes can then take place over a stream of pulse sequences to augment the SNR further. Once all this has been carried out, detection of

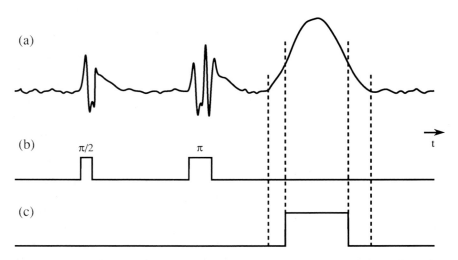

Fig. 17.5   Use of a linear gate to eliminate noise and interference. Here a linear gate (c) is used to eliminate noise and the remanent effects of the microwave pulses (b). In practice a gating pulse may be used which is narrower than the width of the echo signal, so as to optimize the SNR eventually obtained from the incoming signal (a).

electron spin-echoes has the advantage of providing a strong signal well away from the pulses that initiate it, so virtually ideal noise conditions should exist at the detector.

In the end the purpose of spin-echo detection systems is to find more about the magnetic ions or the crystals that contain them. Thus it is necessary to change some variable such as magnetic field or temperature or inter-pulse delay time τ in order to obtain the required data. (Note that the absolute height of the echo holds very little information, since it may not be known exactly how many magnetic ions are in the crystal sample – it is often only *comparative* data that are of value to the physicist.) Thus the boxcar detector will output an average signal level that will vary as the magnetic field or temperature or τ is varied, and this variation can conveniently be drawn out on a chart-recorder.

Next, we consider whether the SNR could be improved further. In fact, we can see that the low-level parts of the echo signal may add very little to the integrated signal, but the noise that is included over this time will be as strong as that arising with the stronger parts of the signal. It will therefore be better to set the linear gate to exclude a small part of the echo and the noise that accompanies it (this is in exact parallel with the situation in Figure 16.4). However, if we weight down this part of the signal, it will provide a useful component to the integrated signal, and the noise it adds will be low enough not to require it to be rejected entirely. Clearly, what we need here is a matched filter, which will weight the high- and low-level signal components in exactly the correct way so that useful signal is not ignored while noise is suppressed as far as possible. The matched filter can be implemented *either* as an especially sophisticated linear gate *or* as a suitably shaped i.f.* amplifier frequency response profile (which should be the Fourier transform of the ideal matched filter time-response function). The parallel with the radar receiver design is particularly noticeable, or even rather startling, considering the underlying differences in the experimental conditions! However, such occurrences are not in the end accidental, but reflect important underlying principles in the ways in which signal recovery can be performed. In fact, the key to the similarities of the two systems is the time-separation of the engineered cause (a pulse or pair of pulses) and the resulting physical effect (an echo signal).

It is interesting to note that spin-echoes can also be produced by nuclear spins in a solid. In that case similar magnetic fields are used, but the resonant frequencies are normally at much lower frequencies, in the range 10–50 MHz, and rather different methods are used to detect the echoes. Again, it is worth emphasizing that the spin-echo approach is especially valuable in that it

---

* See footnote in Section 16.3.

engineers a response from the magnetic spin system at a time when there are no sources of interference from the oscillating magnetic fields that interact with the spins: hence the SNR can be close to its ideal value.

Finally, echoes (called *photon echoes*) have also been produced by pairs of laser pulses applied to crystals such as ruby. We shall not discuss this topic here, but it is interesting to note that echoes are not an isolated and obscure phenomenon, but arise as a consequence of the non-linear responses of various substances to electromagnetic fields.

17.4 Optimizing the electron spin-echo system

At this point it is worth examining in more detail the spin-echo system outlined above. In particular, we should find how to optimize the SNR by adjusting the inter-echo period T and the time τ between the $\frac{\pi}{2}$- and π-pulses. First, note that as soon as the spins have been brought down to the equatorial plane by the $\frac{\pi}{2}$-pulse, they start relaxing back towards the direction of $\mathbf{H_0}$ with relaxation time T_1, according to equation (17.1). If after a time T the available spins are rotated into the equatorial plane by a further $\frac{\pi}{2}$-pulse, the rotating magnetic moment will now be:

$$M_1 = M_0[1 - \exp(-T/T_1)] \tag{17.2}$$

and this determines the maximum observable size of the echo. However, as we have already seen, while in the equatorial plane the spins dephase irreversibly with relaxation time T_2, so that the echo amplitude is reduced according to:

$$M = M_1 \exp(-2\tau/T_2) \tag{17.3}$$

Therefore, in the general case where $\tau > 0$ the echo size will be given by:

$$M = M_0 \exp(-2\tau/T_2)[1 - \exp(-T/T_1)] \tag{17.4}$$

The repetition rate $1/T$ determines the number of echoes that are summed* in the total observation time t_{obs} and hence affects both the total signal and the total noise power:

$$S = (t_{obs}/T)M \tag{17.5}$$

---

* Note that equation (17.4) gives the steady-state echo signal only if all the phase-memory is lost between echo sequences, i.e. only if $T_2 \ll T$.

$$N^2 = \lambda(t_{obs}/T) \qquad (17.6)$$

λ being the basic noise power during the times when the detector is actually observing echo signals.

We now have to optimize the (voltage) SNR:

$$\rho = S/N = (t_{obs}/T\lambda)^{1/2} M_0 \exp(-2\tau/T_2)[1 - \exp(-T/T_1)] \qquad (17.7)$$

It will be clear that it is best to minimize τ, so that $\exp(-2\tau/T_2)$ is as close to unity as possible. How well this may be achieved depends on the extent to which breakthrough from the powerful microwave pulses can be eliminated by PIN diodes or other devices. In practice, useful results can often be obtained with values of τ as small as 250 ns. We now consider adjustment of T. Differentiating ρ gives:

$$\begin{aligned} d\rho/dT = (t_{obs}/\lambda)^{1/2} M_0 \exp(-2\tau/T_2) \\ \times \{[-1 - \exp(-T/T_1)]/2T^{3/2} + [\exp(T/T_1)]/T_1 T^{1/2}\} \qquad (17.8) \end{aligned}$$

This expression is zero when:

$$T_1[\exp(T/T_1) - 1] = 2T \qquad (17.9)$$

Writing $u = T/T_1$ we get:

$$e^u = 1 + 2u \qquad (17.10)$$

$$\therefore \quad u = \ln(1 + 2u) \qquad (17.11)$$

We can solve this equation by iteration, starting with $u = 1$, obtaining finally:

$$T/T_1 = u = 1.256 \qquad (17.12)$$

The ρ v. T optimization curve is shown in Figure 17.6, and after an initially very rapid increase ($d\rho/dT = \infty$ at $T = 0$), it reaches the maximum indicated above and finally falls to zero as $T \to \infty$.

The above calculation shows how to optimize the SNR when the size of the echo is being observed and T, τ do not need to be varied to get extra information from the system. This represents an experiment in which the size of the echo is monitored as magnetic field H_0 is swept through a range of values, so as to search for electron resonances. However, there are other experiments that we might wish to perform. For example it might be useful to sit on a magnetic resonance line and measure the decay of the signal (the

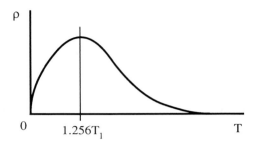

Fig. 17.6   Optimizing spin-echo sensitivity.

echo) as T, or τ, is varied (see for example Figure 17.4), since, by equation
(17.4), this will give the value of T_1, or T_2. In the former case, the above
theory shows that SNR will be optimized if τ is kept as low as possible, while
in the latter case the optimization again clearly leads to $T = 1.256T_1$.

17.5   Use of electron spin-echoes for ENDOR

Another technique that employs electron spin-echoes is called ENDOR
(see Section 13.4). This involves looking for nuclear resonances by sitting
on an electron resonance line. The technique is similar to that used when
measuring T_2. However, instead of varying τ, we vary the frequency of a
further oscillating magnetic field H_N in an effort to stimulate the resonances
of any nuclei in the vicinity of and interacting with the electrons. An obvious
way to do this is to apply the "nuclear" field H_N during the two periods τ;
then nuclear resonances interfere with the formation of the echo from the
electron spins, reducing its size – and the nuclear resonance spectrum is
drawn out on the chart recorder that monitors the electron spin-echo signal.
   Unfortunately, this approach to ENDOR does not work well since T_2 is
too short. This has two effects. The first is that insufficient numbers of nuclear
spins are flipped by H_N in times of the order of T_2, and sensitivity therefore
suffers. The second is that those that are flipped have very broadened
resonance lines since the "nuclear" frequency f_N is not well specified by a
pulse of so short a duration (the Fourier transform of a pulse of duration
$\sim T_2$ has width $\sim 1/T_2$ which is frequently much wider than the width of a
nuclear resonance line, e.g. $T_2 \approx 3\,\mu s$, giving $1/T_2 \approx 300\,kHz$, compared with
a nuclear line width $\sim 30\,kHz$).
   An alternative technique that has been found to solve this problem is the
following. Each sequence consists not just of a pair of electron spin-echo

pulses and a nuclear frequency pulse, but an initial π-pulse, then a long nuclear frequency pulse and finally a pair of electron spin-echo pulses (Figure 17.7). The initial π-pulse "burns" a long-lasting hole in the electron resonance line, which is gradually filled in by spin–spin interactions ($T_1^\dagger$ processes*) and by nuclear resonances (Figure 17.8). Thus the size of the final echo (which would have been greatly reduced by the initial π-pulse) is increased towards its normal value by *both* of these effects (Figure 17.9). If we monitor the size of the echo in the absence of nuclear resonances, we find a signal of magnitude:

$$S = S_0[1 - 2\exp(-t/T_1^\dagger)] \tag{17.13}$$

Here we are assuming (a) that $T_1 \gg T_1^\dagger$, and (b) that the repetition time T between the sequences described above is set at $1.256T_1$ as discussed earlier. If there is a *very* strong nuclear field H_N, then S will be increased immediately from $-S_0$ to zero (it will not go immediately to $+S_0$ unless there is a nuclear π-pulse: normally, the best that can be achieved is to equalize the populations

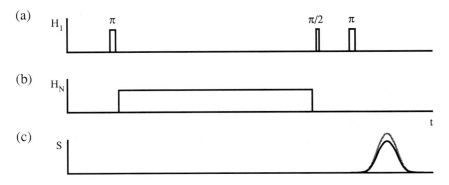

Fig. 17.7 Pulse sequence for performing ENDOR. Here, a π-pulse is performed to "burn a hole" in the electron resonance line (Figure 17.8), then a nuclear frequency H_N pulse is applied, and finally, a standard $(\pi/2, \pi)$-sequence is applied to form an electron spin-echo signal: monitoring this as the nuclear frequency is varied gives the pulse-ENDOR signals. Note that ENDOR here causes an *increase* in the magnitude of the echo signal (see grey curve).

---

* These are considerably slower than T_2 processes because the spins have large energy differences when aligned parallel or anti-parallel to $\mathbf{H_0}$, and are therefore not on "speaking terms": however, these $T_1^\dagger$ processes are generally faster than T_1 processes. The normal ordering is: $T_1 > T_1^\dagger \gg T_2 > T_2^\dagger$.

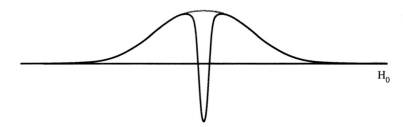

Fig. 17.8   Hole burnt in the EPR line by a π-pulse. A π-pulse inverts the spins at the centre of the electron resonance, but has little effect for far-off parts of the resonance where $\Delta H_0 \gg H_1$. As a result it appears that "a hole has been burnt" in the electron resonance line.

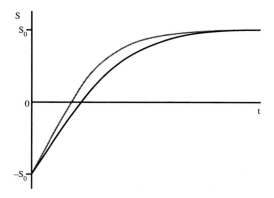

Fig. 17.9   Recovery of spin system after application of a π-pulse. This figure shows the recovery of the spin system after the application of a π-pulse: the black curve shows the recovery resulting from $T_1^\dagger$ relaxation processes alone, and the grey curve shows the increased rate of recovery when nuclear resonance occurs.

of the relevant two nuclear energy levels). This would give a maximum sensitivity of *half* the available electron resonance signal.

   In practice, ENDOR sensitivity tends to be rather low, because of the difficulty of designing an environment at the sample that provides strong H_1 and H_N. Here it will be useful to assume that H_N has the effect of modifying the $T_1^\dagger$ relaxation time very slightly, thereby changing the echo signal from S to S'. To optimize the effect, we examine how much a small change in $T_1^\dagger$ changes S:

$$dS/dT_1^\dagger = -(2S_0 t/T_1^{\dagger 2})\exp(-t/T_1^\dagger) \qquad (17.14)$$

The change in S will be a maximum after a time $T_N = t$, where $d(dS/dT_1^\dagger)/dt = 0$. Now

$$\frac{d(dS/dT_1^\dagger)}{dt} = -(2S_0/T_1^{\dagger 2})[\exp(-t/T_1^\dagger) - (t/T_1^\dagger)\exp(-t/T_1^\dagger)] \quad (17.15)$$

Hence the condition for optimum ENDOR sensitivity is simply $T_N = T_1^\dagger$. Note that in this case we are optimizing $S_N = (S' - S)$ – which is a signal – and ignoring noise. Oddly, this is entirely valid, since the noise is not affected by variation of T_N – so long as the repetition time T is kept constant at $1.256T_1$ as assumed above: i.e. in this case optimizing the signal also optimizes the SNR.

Note that this analysis carries many simplifying assumptions that cannot be discussed in detail here, such as what happens when T_1 is not much greater than $T_1^\dagger$. However, the calculations presented above should give something of the flavour of the optimizations that arise in this sort of work.

17.5.1 Eliminating drift

Finally, we consider what happens if the nuclear resonance field gradually changes in amplitude and causes a variable heating effect which affects the echo size. Drifts from this cause can be virtually eliminated by alternating H_N between off and on, on alternate pulse-echo sequences: then we can proceed *either* by using two boxcar detectors and subtracting their outputs, *or* by subtracting the inputs to a single boxcar detector – the latter approach being preferable for more accurate cancellation. A better alternative than 100% amplitude modulation is to frequency-modulate the nuclear frequency f_N using a square-wave source, and to subtract the two outputs. Then, on approaching a nuclear resonance, the echo signal first increases, and then goes negative, before returning finally to zero – i.e. we have a derivative of the original absorption resonance line (cf. Figure 13.13). In some circumstances, an even better alternative is to *doubly* differentiate the resonance by using a bidirectional square-wave consisting of a sequence of four slightly different values of f_N, namely $f_{N1} = f_N$, $f_{N2} = f_N - \delta f_N$, $f_{N3} = f_N$, $f_{N4} = f_N + \delta f_N$ (Figure 17.10). If the corresponding signal outputs are respectively S_{N1}, S_{N2}, S_{N3}, S_{N4}, then we take the final signal as:

$$S_N = S_{N1} - S_{N2} + S_{N3} - S_{N4} \quad (17.16)$$

Suppose now that the S_{Ni} vary with f_{Ni} as:

$$S_{Ni} = a + bf_{Ni} + cf_{Ni}^2 \quad (17.17)$$

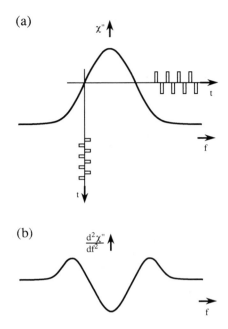

Fig. 17.10 Drift cancellation by double differentiation of ENDOR signals. (a) The ENDOR absorption signal χ'' being detected using bidirectional square-wave frequency modulation: with suitable analysis (see text) this results in the doubly differentiated signal, shown in (b).

Then we obtain:

$$
\begin{aligned}
S_N &= 2[a + bf_N + cf_N{}^2] - [a + b(f_N - \delta f_N) + c(f_N - \delta f_N)^2] \\
&\quad - [a + b(f_N + \delta f_N) + c(f_N + \delta f_N)^2] \\
&= -2c\delta f_N{}^2
\end{aligned}
\tag{17.18}
$$

Thus the constant and linear variations of S_{Ni} with frequency have been cancelled out: i.e. any drift arising from heating or other effects has been cancelled out to second order. Furthermore, the second-order effects can also be cut down by reducing δf_N, though eventually this reduces the magnitudes of the resonances themselves. Clearly the cost of this form of signal processing is to present the original absorption signal as a doubly-differentiated resonance line with one main and two smaller opposite-polarity peaks (Figure 17.10). In most cases where resonance lines are fairly well separated this does not give rise to any significant problems.

17.5.2 Discussion

Overall, pulsed ENDOR is much more systematic to set up and more reliable in principle than the c.w. form of ENDOR described in Section 13.4. There are two reasons for this. The first is that the signal is detected at instants when there is no possible interfering radiation in the equipment; and the second is that the very act of performing the detection involves precise knowledge of most of the relevant relaxation times, so optimization is intrinsic to the process of detection (of course we still have the problem that it will not be known how to optimize a *specific* spectroscopic line until it is actually observed!). In addition, it should be noted that there are more ways to set up and adjust a pulsed ENDOR experiment than the corresponding c.w. experiment, thereby giving greater freedom both to optimize SNR and to interrogate specific nuclei. However, the *basic* sensitivities for c.w. and pulsed ENDOR must be similar, since the magnetization energies of the same numbers of electron and nuclear spins have to be detected against similar noise backgrounds. (Much the same comments apply to c.w. and pulsed EPR or NMR, but with ENDOR applications spectroscopic sensitivity is often very much at the limits of detectability, so there is greater need to optimize the detection system.)

17.6 Magnetic resonance imaging

Like computerized tomography and ultrasound imaging, the purpose of magnetic resonance imaging (MRI) is to map the internal organs of the human body with a view to aiding diagnosis of diseases such as cancer. The possibility of using NMR for this purpose arose as recently as 1973, and since then it has advanced dramatically, with the production of several commercial instruments and numerous special techniques. It will not be possible to describe MRI in detail here, but it will be useful to state some basic principles and to place the methods within the context of the subject matter of the chapter as a whole.

MRI is an adaptation of NMR to probe the (many) hydrogen nuclei in the human body, and to characterize their chemical sites sufficiently to construct useful maps, or "slices", at carefully chosen angles and positions. Since spatial information is the main requirement of MRI, some trick is required to differentiate and record the signals from the separate nuclear locations. This is achieved by applying time-switched combinations of magnetic field gradients, in two dimensions within the chosen slice, and then arranging for nuclear spin-echoes to appear in a carefully controlled manner, so that they give information about the various locations.

First, to select a suitable slice, which we shall assume is to be perpendicular to the z-axis, a strong H_0 field with a field gradient G_z is applied to the sample along this direction. Then a $\frac{\pi}{2}$-pulse is applied to the sample, whose oscillating component is resonant only with the nuclear spins in the chosen slice (by now differentiated from other slices by G_z).

Next, the field gradient G_z is removed, and a field gradient G_x is applied: at this point a free induction decay is initiated by G_x. A little later, this field gradient is reversed to give a *gradient-recalled echo* (GRE) which arises not from a later π-pulse, but from the refocusing due to inversion of G_x. On its own, the time-profile of such an echo will represent the spatial frequency spectrum in the x-direction, for the chosen slice*. Clearly, by applying a time-switched field gradient G_y, we can also get an echo profile that represents the spatial frequency in the y-direction, and by various time-switched combinations of field gradients, a 2-D echo signal can be built up which represents the complete (spatial) Fourier transform of the slice. Performing the 2-D inverse Fourier transform will thus recover the intensity map of the chosen slice.

Applying the various field gradients and building up an adequate 2-D Fourier transform of the slice can clearly take considerable time – sometimes several seconds – and this causes problems as patients breathe or their internal organs move. The technique known as echo-planar imaging (EPI) has been devised to overcome these problems, without too much loss of SNR. In this case a sequence of some 256 echoes is obtained in a single echo sequence (i.e. without waiting each time for a T_1 decay to occur): this interesting possibility arises since any echo can be restimulated by a further π-pulse (or field-gradient reversal), in a very long sequence that is limited only by the T_2 relaxation time (see last paragraph in Section 17.2). With T_2 values of 50–100 ms, enough echoes are obtained to give the required spatial resolution in the chosen slice. This technique shows promise for cardiology, where it would be valuable to be able to study heart-beating effects dynamically in any chosen section, at several frames per second.

A very recent application of MRI has been to brain research, to ascertain which areas of the brain are engaged in thinking activities. During thought processes, active areas of the brain absorb more oxygen, and thus de-oxygenated haemoglobin, which is paramagnetic, increases in concentration.

---

* There is plenty of scope in this problem to be muddled by the different variables. First, note that constant G_x makes x proportional to H_x which, in turn, is proportional to the local nuclear resonance frequency f. Now the echo is observed as a time development. Hence the echo profile is also a spatial frequency pattern, with independent variable f_x. To find the x-variation we therefore need to find the (inverse) Fourier transform of the echo profile. The potential muddle arises as f_x is analogous to t, not f.

It has now been found that MRI is able to map the local changes in magnetic field produced by deoxyhaemoglobin – a technique known as blood oxygen level dependence (BOLD) imaging.

17.7 Summary

This chapter has been concerned with recovery of signals from magnetic spin systems that are subjected to various pulse sequences. It was necessary to devote a fair amount of space to the description of the basic physical phenomena, and specifically to show how the magnetic spins dephase after a $\frac{\pi}{2}$-pulse and how they are refocused to yield an echo after a π-pulse. (Readers who are less interested in the physics of echo systems may like to regard these experiments as rather elaborate forms of radar in which pairs of pulses are transmitted instead of single pulses.) There is some advantage to be gained from employing this pulsed form of detection instead of the c.w. approach (see Chapter 13), since (as in radar) the signal appears at a point in time well away from the pulses that initiate it, so virtually ideal noise conditions can exist at the detector. There is considerable scope for optimizing the detection of spin-echoes, and the chapter covers some of these aspects of the problem. In addition, it shows how drift in pulsed spin-echo ENDOR spectroscopy can be cut down by using bidirectional square-wave frequency modulation to doubly differentiate the nuclear resonances. Pulsed detection methods are found to offer other advantages over c.w. methods as measurement of relevant relaxation times is implicit, so optimization is essentially intrinsic to the process of detection. Finally, magnetic resonance imaging is an excellent example of an initially purely academic idea (the spin-echo) being turned to the benefit of mankind. Although it has not reached the end of its development, this tool has already undoubtedly saved many lives, and demonstrates the unity of science and engineering.

17.8 Bibliography

Magnetic spin-echo systems are EPR and NMR systems that are operated in pulsed mode. Thus the references on EPR and NMR given in Chapter 13 still have a general relevance: these include the books by Orton (1968), Abragam (1961), Abragam and Bleaney (1970), the paper on c.w. ENDOR by Davies and Hurrell (1968), and the review by Baker, Davies and Reddy (1972). To these must be added the book on electron spin relaxation measurement by Standley and Vaughan (1969) which includes details of pulsed EPR techniques. Work on pulsed ENDOR was initiated by Mims (1965), who evolved a very cunning technique for burning a complex hole in the electron resonance line, and this was later followed by the rather simpler method developed by the author (Davies, 1974) and described in the text. It turns out that Davies-ENDOR can have advantages over Mims-ENDOR because it is not subject to certain blind-spots in the resonance spectra. For recent appraisals of the two techniques, see Grupp and Mehring (1990) and Gemperle and Schweiger (1991).

NMR spin-echoes have long been known and understood (see, for example, Abragam, 1961). Recently they have been the subject of detailed studies (e.g. Sebastiani and Barone, 1991), because of their relevance to magnetic resonance imaging. For an up-to-date account of magnetic resonance imaging techniques, see Stehling et al. (1991); the reader may find it useful to refer to the original paper on EPI by Mansfield and Pykett (1978), the subsequent publication by Johnson et al. (1983), and the volume by Mansfield and Morris (1982). Perhaps surprisingly, spin-echoes were first reported as early as 1950 by Hahn, for NMR, and the multiple π-pulse echo sequence mentioned in Section 17.2 was developed soon after by Carr and Purcell (1954); spin-echoes were extensively developed by Mims et al., for EPR, in an important paper (1961) nearly a decade later. The concept of echoes also appears in the very different area of photon echoes – see the original paper by Abella et al. (1966) and the early popular article by Hartmann (1968). For a recent, incisive account of spin-echoes and photon echoes in a broader context (and leading onto interesting topics such as self-induced transparency), the reader is referred to Macomber (1976).

17.9 Problems

1. Show that a FID is initially linear in shape if there is no inhomogeneous broadening, and initially quadratic in shape if the broadening is almost entirely inhomogeneous. Hence deduce that a spin-echo will always have a rounded peak rather than a pointed peak.
2. In a certain spin-echo experiment, it is decided to average over n echoes *within* the FID by employing n π-pulses after the initial $\frac{\pi}{2}$-pulse. Obtain a formula for the gain in SNR relative to the case of $n=1$. Show that straight averaging of the n echoes results in an optimum for a value of n less than the maximum possible. Assume that $T_2 \gg T_2^\dagger$, T_{exp}, where T_{exp} is the minimum time between echoes that can be produced by the equipment.

3. In a pulse-ENDOR experiment, determine whether the optimum moment for applying the observing echo sequence gives a normal or inverted echo in the absence of a nuclear resonance. If instead the echo sequence is set up so that the echo has zero amplitude, by what factor will the ENDOR SNR be reduced below its optimal value? (Adjusting the echo for zero amplitude might be a useful rule-of-thumb for setting up the system if this gives a SNR that is quite close to optimal.)

4. Determine the optimal modulation depth (δf_N) for detection of a Gaussian ENDOR line of standard deviation σ using the double differentiation method to eliminate drift. (Note that it will be better to ignore the wings of the Gaussian if there is severe drift, but ignoring much more will cut down the signal too far: one way of performing the analysis involves approximating the Gaussian by an inverted parabola.)

5. It is postulated that a spin-echo system can be used to store information for a long time ($\sim T_1^\dagger$) by splitting the π-pulse into two $\frac{\pi}{2}$-pulses. After the usual initial $\frac{\pi}{2}$-pulse and the first half π-pulse, the spins are in a vertical plane; after some time the lateral components of their magnetic moments will have decayed away, but the vertical components (parallel to $\mathbf{H_0}$) will remain for a time $T_1^\dagger$ much greater than T_2. Then the final half π-pulse is applied, bringing the remaining spin components back into the equatorial plane, and an echo appears. Find the loss in amplitude of the spin-echo resulting from decay of the lateral components of spin: ignore the effects of $T_1^\dagger$ decay.

18

Detection of Radio Signals

18.1 Introduction

The detection of radio signals is a very wide subject and space does not permit it to be studied at length in this book. Yet it has a great deal of relevance to the subject of signal recovery. First, like radar, it involves the extraction of very small signals that are embedded in considerable noise. Second, the concepts of radio apply immediately to radar, which after all involves radiation at the uppermost end of the radio frequency scale – so (for example) the techniques of superheterodyne reception are still relevant. Indeed, they are also relevant when EPR, NMR or even optical signals are being detected. Thus it will be valuable to examine some of the problems of radio signal detection: this chapter will concentrate on the principles of various types of radio receiver, and will then embark on a detailed analysis of detection and mixing processes, studying particularly the effects of noise in detection systems.

18.2 Basic problems of signal detection

We start by taking the case of an amplitude-modulated carrier wave and see how the modulation frequency can be extracted. In principle this is easy since all we need to do is to rectify (or "detect") the signal, and smooth it over a number of cycles (Figure 18.1). However, there is a specific problem with this approach when any noise is present. (Note that the inverse square law has a devastating effect on the amplitudes of radio signals, and it is therefore the norm for them to be subject to considerable levels of noise.)

(a)

(b)

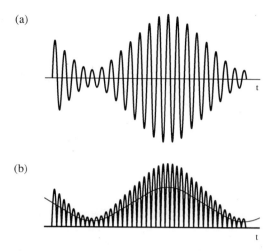

Fig. 18.1 Detection of an amplitude-modulated carrier. Here (a) shows an amplitude modulated carrier, and (b) shows both the detected (full-wave rectified) signal and the smoothed form that provides the demodulated signal.

Suppose first that the signal is much greater than the noise. Then the signal waveform switches the detector diode and the noise is effectively resolved along the signal phasor: this is because the small noise voltage has little effect on the angle of the resultant phasor, but has an additive effect on the signal amplitude (Figure 18.2). Now consider what happens when the noise is much larger than the signal. In that case the detector diode is switched randomly by the noise, and the signal merely modulates the amplitude of the noise. However, since the noise is uncorrelated with the signal, and vice versa, the signal appears as a further noise component on top of the noise. (Indeed, the signal can be shown to add to the noise power! See Section 18.5 for relevant theory of this and other results quoted in the present section.) The important point is that all signal information is essentially lost. The intermediate case is quite complicated and we shall not consider it in detail here. Suffice it to say that there is a highly non-linear signal response function when the signal level is of similar size to the noise voltage, and the characteristics are not at all suitable for radio detection purposes.

A further problem also arises in this context, and that is the behaviour of signal diodes at low voltage levels. When signal voltages are low – and certainly when they are small fractions of a volt, it is difficult to find electronic devices such as diodes that have a sufficiently non-linear effect to rectify them effectively. For example, junction diodes obey the exponential law, which is

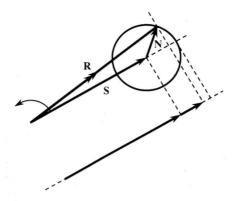

Fig. 18.2   Effect of noise on an a.c. signal. Here, **S** represents the phasor of the signal waveform, **N** is the noise phasor, and **R** is the resultant. Note that if |**N**| is small compared with |**S**|, then **N** has little effect on the angle of **R**, but has an additive effect on the signal amplitude |**S**|. However, when |**N**| is of the order of |**S**|, or larger, the situation is much more complex (see text).

a smoothly varying function near the origin, and below signal levels ~ 0.1 V the non-linear effect is too small to give adequate detection. The Schottky barrier diode and the back diode have the capability of working at slightly lower voltages, but do not remove the basic problem. This is therefore another problem that has to be solved if modulated signals are to be decoded with adequate SNR.

18.3   The superheterodyne receiver

Superheterodyne detection aims to overcome these problems. First, the detector diodes are switched by a strong waveform that is dominant over both the incoming signal and any accompanying noise voltages. Second, the switching or "local oscillator" waveform is in general a sine-wave of similar frequency to the carrier wave. Thus the local oscillator waveform beats with the signal waveform and produces sum and difference tones. The sum tones are eliminated by a low-pass filter, leaving just the difference tone, which is at a so-called *intermediate frequency* (i.f.), much lower than the carrier frequency, which can be amplified with relative ease. However, the amplitude of the i.f. signal is fairly close to that of the original carrier wave (assuming that an ideal switching mixer is used – see Section 18.5.1). Thus all that has so far been achieved is to institute a change to a lower, more convenient frequency.

There is, however, another important effect. This is that all incoming carrier frequencies that are to be detected are converted to the same i.f. so that they can be handled in a similar manner. This has the advantage that wide-band r.f. amplifiers are not needed, and also that the i.f. amplifier passband can be accurately shaped so as to let through the frequency-changed carrier with all necessary sideband information: hence the *fidelity* of the signal is significantly improved, and also the tuning *selectivity* of the receiver is vastly improved (there is always a tradeoff between these two parameters, but with a superheterodyne receiver the tradeoff can be controlled overtly and brought close to the ideal). Furthermore, the gain of an i.f. amplifier can be made much higher since it only has to cope with a single narrow frequency band. Finally, the large gain means that the signal can be brought up to levels that can switch a final detector adequately, while noise and clutter are minimized by the reduction in bandwidth of the passband. Thus the final detector circuit is able to operate under near-ideal conditions.

We have called the frequency changing stage a detector as it has a non-linear element (usually a diode), but it is more often called the *mixer* or *first detector*. The final detector is simply called *the detector* or otherwise the *second detector*.

The superheterodyne receiver is considerably more complex than the older simple detector stage, as it needs local oscillator and mixer stages (Figure 18.3): the i.f. amplifier replaces the tuned r.f. (t.r.f.) amplifier that would otherwise be present (note, however, that many "superhet" receivers have an initial t.r.f. stage of moderate gain as well as an i.f. stage, the purpose being to prevent the mixer from being overloaded by currently irrelevant signals). Although, as indicated above, the superhet solves a number of problems, it is still unable to solve the "threshold" problem referred to earlier, whereby signals which are still below the noise level on reaching the second detector cannot be detected. The main possibility of solving such problems lies with synchronous detection systems, which are discussed below.

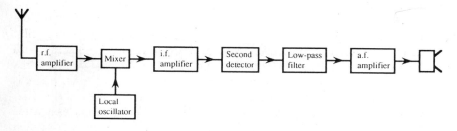

Fig. 18.3 Block diagram of superheterodyne receiver.

18.4   Synchronous detection systems

Synchronous detection can be considered as a special case of superheterodyne detection in which the local oscillator frequency is identical to that of the expected signal. The mixer diodes are again switched positively and the signal is detected coherently irrespective of the strength or phase of the noise waveform (assuming the latter is at a fairly low level).

One approach to synchronous detection is the *homodyne* receiver (Figure 18.4). This operates by amplifying the carrier and "limiting" it to cut out amplitude modulation or any variations in amplitude due to noise, thereby creating a local oscillator waveform of the same frequency as the carrier. Unfortunately, the homodyne is restricted in that severe fading of the carrier will temporarily eliminate the local oscillator frequency and the receiver will not operate. This problem can partially be overcome by using the amplitude-limited carrier to drive a high-Q tuned circuit which will maintain the oscillations for some time. Even so, strong fading will still prevent the circuit from working, and in any case if the incoming carrier is below noise level the whole strategy becomes inoperable.

We are then driven to a second approach in which a separate oscillator is employed to act as a local oscillator. This oscillator is voltage-controlled and is inserted into a feedback loop which is designed to maintain the oscillation at the same frequency and phase as that of the incoming carrier. The whole device is called a *phase-locked loop* (p.l.l.). This contains not only the voltage-controlled oscillator (v.c.o.) but also a *phase comparator*, a d.c. amplifier and a low-pass filter. We describe the characteristics of a phase comparator below. Meanwhile, note that the resulting *synchrodyne* receiver employs a mixer, low-pass filter, 90° phase-shift circuit, and d.c. amplifier (or a.f. amplifier) (Figure 18.5). Perhaps the most characteristic difference between the synchrodyne and the superhet is that (as intended) the i.f. is now centred at d.c., so a *single* low-pass filter can be used for signal selection instead of a number of tuned circuits.

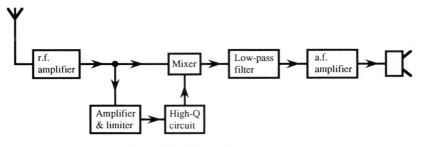

Fig. 18.4   Homodyne receiver.

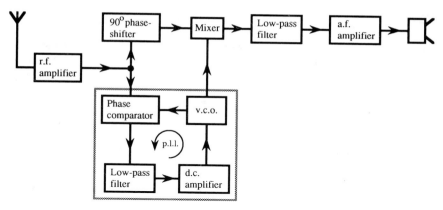

Fig. 18.5 Synchrodyne receiver. In this circuit, note that four of the blocks can be considered to form a phase-locked loop (p.l.l.), which is outlined in grey. The v.c.o. is a voltage-controlled oscillator.

Clearly, the p.l.l. is a particularly vital component of a synchrodyne receiver, while the phase comparator is a most vital sub-component. A *true* phase comparator is very similar to a phase-sensitive detector, but is designed to be insensitive to signal amplitude. This is achieved by amplifying and limiting its two input signals. Ideally, we then have two square-waves which can be fed into a normal p.s.d., whose average output will then be proportional to the period of time both waveforms are positive together – i.e. output will be proportional to $(1-2|\delta|/\pi)$, where δ is the phase difference. Thus the output will range from $+1$ when $\delta=0$ to -1 when $\delta=\pm\pi$, and will be zero when $\delta=\pm\pi/2$. In fact a variety of devices can perform this function, and we might alternatively design the device using an analogue multiplier. On the other hand the square-wave inputs are essentially logic inputs, and as such we could use a *coincidence* gate (an inverted *exclusive or* gate – see Table 18.1) together with a time-constant or other averaging circuit to process them – see the waveforms in Figure 18.6.

It is of interest that a phase comparator can only give output levels between -1 and $+1$ corresponding to phase differences between $-\pi$ and $+\pi$. If phase differences outside these limits arise, they are curtailed appropriately. This is relevant since the p.l.l. can only stabilize frequencies over a certain capture (or *lock*) range, and if frequencies drift outside this range the feedback loop will no longer be able to control them. Note also that the loop gain is important in determining *both* the capture range *and* the degree of stabilization provided (i.e. the amount that a small frequency or voltage perturbation will be reduced by the feedback loop). An important factor in setting up a p.l.l. is to adjust the frequency carefully before closing the

Table 18.1 Logic for a phase comparator. This table shows the logic required for a phase comparator: S is the signal input, R the reference input, $\odot$ the *coincidence* function, and $\oplus$ the *exclusive or* function. The averaged coincidence output is the function envisaged for a phase comparator. However, note that the logic levels at the output of the coincidence gate will have to be re-interpreted $(1 \to 1; 0 \to -1)$ to give the stated phase-comparator outputs (see text).

S	R	$\odot$	$\oplus$
0	0	1	0
0	1	0	1
1	0	0	1
1	1	1	0

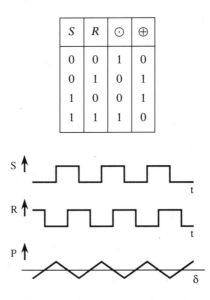

Fig. 18.6 Specification of a phase comparator. A phase comparator accepts two square-wave inputs, S and R, and gives an output P which varies as the phase-difference δ between them varies, being zero when the two waveforms are in quadrature as shown.

feedback loop – i.e. close the loop to stabilize the frequency only after careful signal selection. Note also that when a p.l.l. does lock, the phase difference δ is maintained very close to $\pi/2$, so that zero signal amplitude is available at the output of the v.c.o.: this explains why a 90° phase-shifting circuit is required in the receiver of Figure 18.5. Notice also that the position where δ is very close to $-\pi/2$ is not a position of stable equilibrium – clearly, the phase is here 180° different from that at the stable position, so the feedback is positive rather than negative.

When a phase comparator is used in a radio receiver (whose input signal may be below noise), it is not sensible to limit the input amplitude as suggested above: in fact only the input from the v.c.o. should be limited. Thus we must use not a true phase comparator but a p.s.d.

Although the synchrodyne is elegant and appealing, it has by no means replaced the superheterodyne in radio reception because it operates around d.c.: this means that flicker $(1/f)$ noise will be relatively high – it is one of the strengths of the superhet that it is able to avoid flicker noise. It also means that combining the techniques will be the basis for a good solution: this conclusion is also supported by the fact that p.l.l.s often have a limited frequency range (typically less than 100 MHz), so the frequency of the incoming signal may have to be reduced before a p.l.l. can be used.

18.5   Mixers and detectors

Mixers and detectors perform related types of function, and are perhaps best considered together. Essentially, they both engineer frequency changes, in the first case from a signal frequency to another chosen frequency, and in the second case down to d.c. Indeed, in the latter case, the change could be made by a mixer using a local oscillator whose frequency is identical to that of the signal: however, such synchronous detection schemes will not be our prime interest in this section.

Mixers and detectors are also similar in the manner in which their functions are brought about. Non-linear devices have to be used for this purpose. Two main types of non-linear device should be considered – square-law and linear detectors (or mixers), the latter being so named because they exhibit a piecewise linear I–V characteristic with break-point (ideally) at the origin (Figure 18.7). These two types of device are idealizations, but they are quite well approximated by certain practical devices over certain voltage and current ranges. However, the exponential law of the junction diode is not an especially good approximation to either type, though near the origin, for small signals, it may be taken to have an almost square-law characteristic.

18.5.1   Modelling the mixing process

The square-law characteristic provides a useful mathematical model of the process of mixing. For suppose we have two waveforms, of which one can be taken to be the input signal and the other can be taken as the local oscillator or reference waveform:

$$V_s = V_{s0}\cos(\omega_s t + \delta_s) \tag{18.1}$$

$$V_r = V_{r0}\cos(\omega_r t + \delta_r) \tag{18.2}$$

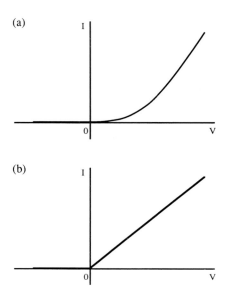

Fig. 18.7 The two main types of non-linear detector. (a) The the characteristic of a square-law detector, and (b) that of a "linear" detector. Note that detectors are necessarily non-linear devices, and that in (b) should properly be called *piecewise-linear*.

Then the result of adding these waveforms and passing them through a square-law device will be proportional to:

$$V_o = \gamma V_{so}^2 \cos^2(\omega_s t + \delta_s) + 2\gamma V_{so} V_{ro} \cos(\omega_s t + \delta_s) \cos(\omega_r t + \delta_r)$$
$$+ \gamma V_{ro}^2 \cos^2(\omega_r t + \delta_r)$$
$$= \tfrac{1}{2}\gamma V_{so}^2 [1 + \cos(2\omega_s t + 2\delta_s)] + \tfrac{1}{2}\gamma V_{ro}^2 [1 + \cos(2\omega_r t + 2\delta_r)]$$
$$+ \gamma V_{so} V_{ro}\{\cos[(\omega_s - \omega_r)t + (\delta_s - \delta_r)] + \cos[(\omega_s + \omega_r)t + (\delta_s + \delta_r)]\}$$

$$(18.3)$$

where γ is a small numerical coefficient of dimensions V^{-1}. Thus we get sum and difference beat tones, and it is usual to eliminate the higher-frequency sum tone, and other tones at double the fundamental frequencies. This leaves the lower-frequency heterodyne signal, which is the one normally used in superheterodyne receivers.

Now let us consider the possible use of linear detectors for mixing. Take an ideal linear detector. The characteristic for this can be approximated over the range $-1 \leqslant V \leqslant 1$ as the series:

$$V_o \approx 0.031 + 0.500V + 1.580V^2 - 3.758V^4 + 4.803V^6 - 2.180V^8 \quad (18.4)$$

Notice that there are no odd power terms after the first, this being explainable since the $0.5V$ term is the mean of the two main piecewise linear components, and the remaining terms can be regarded as correction terms.

Of these terms, the constant term and the odd power term produce no mixing effect, the V^2 term clearly produces a mixing effect, while the higher power terms also have a mixing effect: for example, the V^2 term produces $(\omega_s - \omega_r)$, $(\omega_s + \omega_r)$, $2\omega_s$, $2\omega_r$ and d.c. components, and it is clear that V^4 and higher terms will produce beats between these various frequencies. Overall, the problem here is not *whether* the device can mix two incoming frequencies to produce a difference frequency, but *what proportion* of the input power is converted to the difference frequency instead of to some arbitrary multiple, or sum, or sum plus difference frequency – i.e. we need to measure the *conversion efficiency* η of the mixer. It turns out that η varies with V_r, and assuming that $V_r \gg V_s$, but is still small, then the important term is the square-law term, and the mixer operates as a square-law mixer. However, if η is increased by increasing V_r, then η at first increases and then reaches a maximum: the device has changed from a square-law mixer to a switching mixer, in which the detector diode is switched (i.e. controlled entirely) by the local oscillator waveform. At that point there is no advantage to be gained from increasing V_r further, and the conversion efficiency (amplitude of the difference waveform relative to that of the input signal, V_s) is $\eta = 1/\pi$. The same efficiency applies if a signal is detected synchronously by a local oscillator whose frequency is identical to that of the signal, assuming the phase difference is also maintained at zero: in that case the reference waveform is in fact performing half-wave rectification of the signal, so the usual half-wave output level of V_{so}/π applies.

18.5.2   Modelling detection

We now move on to consider detection. In this case it is the d.c. component that is the important one. The basic equations are very much simpler. We apply the signal waveform to the square-law detector, obtaining merely the term:

$$V_o = \gamma V_{so}^2 \cos^2(\omega_s t + \delta_s) \tag{18.5}$$

which averages to $\frac{1}{2}\gamma V_{so}^2$. On the other hand, for a linear detector, which is a half-wave rectifier, we find the mean (d.c.) output signal level is V_{so}/π. Notice that this is the same as deduced above for a linear detector used as a mixer between a strong reference waveform and the input signal. This is easy to explain since the signal is acting as its own reference and in particular is strongly switching the detector diode. However, this argument also indicates the main weakness in the scheme, since if any noise is present then switching will no longer be controlled exclusively by the signal waveform, and performance will suffer – as we shall now see.

18.5.3   Detection when signals are accompanied by noise

When noise is present, a full analysis of the situation is rather complex, and instead we concentrate here on the case of a linear detector and try to draw some general observations of the SNRs available for this form of detection. First, consider the phasor $\mathbf{V_s}$ for the basic signal V_s and the phasor $\mathbf{V_n}$ for a noise component V_n. These give a resultant phasor $\mathbf{V_t}$ as shown in Figure 18.8: it is the resultant phasor that switches the detector diode, and which would ideally have the same timing and switching capabilities as the signal itself.

When $V_s \gg V_n$, we see that $V_t \approx V_s$, and indeed $\mathbf{V_t} \approx \mathbf{V_s}$. Thus $\mathbf{V_t}$ and $\mathbf{V_s}$ have nearly the same directions, i.e. they rotate with similar timings and therefore $\mathbf{V_t}$ would switch the detector diode almost identically to $\mathbf{V_s}$. However, this does not mean that the noise has no effect. In fact, $\mathbf{V_n}$ modulates the amplitude of $\mathbf{V_s}$: the effect of this is computed by resolving $\mathbf{V_n}$ along $\mathbf{V_s}$ and adding the resulting noise component to V_s. We then find:

$$V_s' \approx V_s + V_n \cos[(\omega_s - \omega_n)t + (\delta_s - \delta_n)] \qquad (18.6)$$

An alternative way of looking at this is to regard the detector as a mixer which mixes the signal and the noise waveforms. In addition, it clearly also mixes the different frequency components of noise with each other, a factor

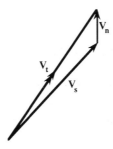

Fig. 18.8   Phasors for signal and noise. Here, the signal $\mathbf{V_s}$ and noise $\mathbf{V_n}$ phasors combine to give a resultant $\mathbf{V_t}$, which in general has different amplitude and phase from those for the original signal.

that has so far been ignored. Returning to the mixing of signal and noise, when $V_s \gg V_n$ we can expect that the result will be a noise-induced signal of size V_n/π, which will be added to the detected signal V_s/π, though the frequencies will be totally different. This effect gives the same result as the formula for V_s' above. A full treatment of this effect involves noticing that the noise boosts the size of V_t, and therefore augments the signal:

$$V_t^2 = [V_s + V_n \cos(\varphi_s - \varphi_n)]^2 + [V_n \sin(\varphi_s - \varphi_n)]^2$$
$$= V_s^2 + 2V_s V_n \cos(\varphi_s - \varphi_n) + V_n^2 \qquad (18.7)$$

so the apparent mean size of V_s is:

$$(V_s)_{\text{eff}} = [V_s^2 + V_n^2]^{1/2} = V_s[1 + V_n^2/V_s^2]^{1/2}$$
$$\approx V_s(1 + V_n^2/2V_s^2) = V_s + V_n^2/2V_s \qquad (18.8)$$

Next, consider what happens when $V_s \ll V_n$. In that case the detector diode will be switched totally randomly by $\mathbf{V_n}$, and $\mathbf{V_s}$ will look like an additional independent random waveform. This means that $\mathbf{V_s}$ will essentially give an additional noise component, and will not be observable as a signal *per se*. The total noise power will thus be:

$$V_n'^2 = V_n^2 + V_s^2 \qquad (18.9)$$

However, there is a remanent detected signal, and it can be estimated as follows: in fact, the resultant signal follows the exact reverse of the situation dealt with above, and the result is obtained by taking the signal part of the resultant:

$$(V_t)_s = [V_n^2 + V_s^2]^{1/2} - V_n = V_n[1 + V_s^2/V_n^2]^{1/2} - V_n$$
$$\approx V_n(1 + V_s^2/2V_n^2) - V_n = V_s^2/2V_n = V_s \times (V_s/2V_n) \qquad (18.10)$$

This shows that the strength of the signal is considerably reduced by the presence of a larger level of noise.

We can summarize these effects as follows:

1. A strong signal is *boosted* by noise.
2. A small signal is *reduced* by the presence of noise.

It is also found that:

3. A strong signal increases the noise by a significant factor relative to the case of no signal.

However, this final effect, not derived above, is quite complex in origin and depends on the type of detector. For example, with a square-law detector, the signal brings the noise into the higher-gain reaches of the characteristic and this has the effect of enhancing the noise. For a complete understanding of this topic the reader should refer to Robinson (1974) and also to Rice (1944, 1945). These references also give more detailed calculations which correct the approximate factors derived above in respect of effects 1 and 2.

18.5.4 Mixing of noise with noise

We now consider the mixing (beating) of noise with noise. Here we recall that noise acts as a multitude of components at the various frequencies within the input bandwidth Δf_0. It will be clear that each of these components can beat with each of the others under the action of the detector diode. Thus we can expect that if the input passband is from $f_0 - \Delta f_0/2$ to $f_0 + \Delta f_0/2$, and the input noise spectrum is white, then the output signal passband will be from 0 to Δf_0, and the output noise spectrum will derate linearly from a maximum at 0 to zero at Δf_0 (the spectrum does not derate to zero at $2\Delta f_0$ because the maximum value of Δf is Δf_0) (see Figure 18.9). This means that the originally flat noise spectrum has become shaped, though the overall bandwidth is unchanged. If in addition, a signal is present, all the noise components will mix with it, thereby producing a more complex noise spectrum. In particular, for a square-law detector, if the signal is at the input passband centre frequency f_0, then the extra signal-induced noise will cut off at $\Delta f_0/2$ (see Figure 18.10). We do not pursue this further as the detailed calculations are very complex, and we have already indicated the main principles that will be of use in practical radio and signal recovery applications. However, it is worth noting that the linear detector gives more

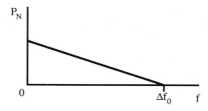

Fig. 18.9 Noise at the output of a square-law detector. This figure shows how the noise at the output of a square-law detector derates to zero at Δf_0, where Δf_0 is the input bandwidth. Here, it is assumed that the detector is subject to pure white noise, and that no signals are present.

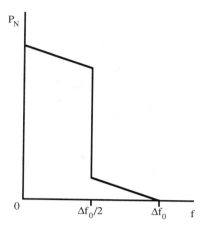

Fig. 18.10 Noise at the output of a square-law detector, when a signal is present. This figure shows how the presence of a signal adds signal-induced noise at the output of a square-law detector (cf. Figure 18.9). It is again assumed that the input noise is white.

complex results, and introduces additional noise components outside the range up to $\Delta f_0/2$, because of the higher harmonics introduced by the later terms in equation (18.4).

Although the output noise bandwidth B_n cannot exceed the input bandwidth $B = \Delta f_0$, it can be quite low if there is a good input SNR: in that case, B_n will be limited entirely by the overt output signal bandwidth b which is determined by the low-pass filter following the detector (it is assumed here that $b \ll B$). But if there is significant noise, the beats between the various input noise components that fall within b at the output contribute to an effective output noise bandwidth B_n that can be greater than b. In general we will have $b \leqslant B_n \leqslant B$. Furthermore, B_n will vary with the input SNR, as indicated above. It is sometimes taken as useful to consider what happens when the output SNR is unity so that the signal is just detectable above noise: in *that* case it has been shown that there is an effective noise bandwidth $B_n = \sqrt{2Bb}$. However, some care must be applied when using this formula, since unity output SNR occurs when the output SNR is proportional not to the input SNR, but to its square (similarly, the output signal is proportional not to the input signal but to its square). Thus the value of B_n given by the above formula is not a rigorously defined bandwidth. However, it is potentially useful in giving an *indication* of the effective noise bandwidth of a system consisting of an amplifier of bandwidth B fed to a detector circuit of output bandwidth b.

18.6   Frequency modulation

Before completing this chapter, some mention should be made of the practically important topic of frequency modulation (f.m.). This is a scheme in which the modulating signal is used to vary the frequency of the carrier, rather than its amplitude. The technique has an advantage over amplitude modulation (a.m.) in that it is a form of coding which is more resistant to noise. Space prevents a full account of the scheme being given here, but a number of observations are in order.

First, the basic f.m. decoder is a device called a *discriminator*, which provides linear frequency-to-voltage conversion. An early type of discriminator employed two *L-C* circuits for this purpose, these being tuned to slightly different frequencies and their detected outputs subtracted to give the required symmetrical response curve. Clearly, a discriminator of this type will also be affected by incidental amplitude modulation. Thus it must be preceded by a *limiter* which will remove any variations in amplitude (Figure 18.11). At its simplest, a limiter may be considered as a high-gain amplifier followed by a resistor feeding a pair of identical Zener diodes, these being placed back-to-back and connected in series to ground: in fact, such a circuit is able to limit l.f. waveforms symmetrically. In practice, more sophisticated circuits must be used at high frequencies. However, some forms of discriminator such as the ratio detector are able to provide ~ 30 dB of limiting, and therefore a separate limiter is not always necessary. Details of the operation of such circuits are beyond the scope of the present volume.

It will be clear from the above discussion that the problem of "fading", which can be severe in a.m. systems, is largely overcome in f.m. systems by the action of the limiter. In addition, limiting is able to remove much of the noise (especially impulse noise) to which radio signals are subject, and the

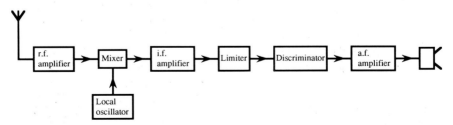

Fig. 18.11   Block diagram of f.m. superheterodyne receiver. This diagram should be compared with that of Figure 18.3 for an a.m. superheterodyne receiver. Note in particular that the second detector of the latter is replaced by a limiter and a discriminator in the f.m. circuit (see text).

quality of f.m. radio reception is therefore considerably better than that of a.m. It is worth noticing that amplitude noise is immediately interpreted as signal by an a.m. detector, whereas in an f.m. receiver, low levels of amplitude noise should change the orientation (i.e. the phase) of the f.m. phasor only by small amounts and will therefore only affect f.m. signals to second order.

To ensure that f.m. systems achieve good noise suppression, it is normal to permit the modulating signal to produce quite large deviations in carrier frequency, the peak frequency deviation Δf typically being 75 kHz for an audio channel of bandwidth $B = 15$ kHz. In that case f.m. provides some 19 dB more protection against noise than an a.m. system with synchronous detection – the relevant formula for the gain in (voltage) SNR being $\sqrt{3}(\Delta f/B)$. However, it must be emphasized that this protection occurs only for low input noise levels. When the noise becomes so high that the phase starts to become randomized (corresponding to an input SNR of about 9 dB), then f.m. has no advantage over synchronously detected a.m. Thus asynchronous a.m. and f.m. systems both suffer from threshold effects, but in broadcast radio reception (where output SNR is normally well over 40 dB), f.m. can be expected to offer considerable advantages over a.m.

A full discussion of f.m. necessarily involves an analysis of the complex sideband structure of f.m. signals, and cannot be entered into here. Suffice it to say that by engineering a wide sideband structure of this type, f.m. signals are coded in a way that makes them readily distinguishable from naturally occurring noise, and ultimately it is this that accounts for the improved reception of f.m. In this respect, f.m. systems exploit the results of Shannon's law (Appendix D, equation (D.5)), which shows much more generally that it is possible to trade bandwidth for an improvement in SNR without change of communication rate. (For f.m. the approach works because, although coding the signal to cover a wider bandwidth initially introduces more noise, final decoding recovers the signal to the narrow bandwidth, while leaving the noise spanning the wide bandwidth; thus low-pass filtering can remove the excess noise.)

18.7 Summary

This chapter has analysed the process of detection of radio signals. In radio, as in radar, signals appear at low level, often deeply submerged in noise, so sophisticated methods have to be used to detect them. One of the main problems is that detector diodes may be switched by noise: this means that a strong local oscillator is required to perform the switching more reliably. If the local oscillator is not at the frequency of the received carrier then the mixer produces an intermediate frequency (i.f.), the receiver then being called a superhet: since the i.f. can be set at a fairly high frequency, flicker noise can be reduced to very low levels. On the other hand, when the local oscillator frequency equals that of the received carrier, as in a synchrodyne receiver, flicker noise is not suppressed. However, the synchrodyne receiver does have the advantage that tuned amplifiers are not required and that a standard phase-locked loop (consisting essentially of a voltage-controlled oscillator and a p.s.d.) will do much of the relevant processing; in addition, it is not subject to any threshold effect.

The chapter also includes a mathematical analysis of the processes of mixing and detection, in the special cases of square-law, linear and switched detectors – the last of these having the best available conversion efficiency. When these detectors are subjected to noise and noisy signals, the situation becomes much more complex. Specific results for square-law and linear detectors are that a strong signal is boosted by noise, while a small signal is reduced by the presence of noise; in addition, a strong signal can increase the noise by a significant factor. Details of these latter aspects are rather complex, and a full discussion is beyond the scope of the present text.

Finally, the reasons why frequency modulation offers improved SNR over amplitude modulation during normal radio reception were outlined, and the existence of a threshold effect in both cases was underlined.

18.8   Bibliography

Radio signal transmission is a well-worn topic and is extensively treated in many books. For example, the superhet dates from the First World War (it was patented by Levy in 1917). However, synchrodyne detection systems arrived much later (around the 1950s). Horowitz and Hill (1989) cover useful ground on phase-locked loops as well as many practical details of superheterodyne receivers and r.f. technology: see also the books on phase-locked loops by Blanchard (1976) and Best (1984). Basic theory of mixing is given by Betts (1970) and by Goodyear (1971); these two volumes also cover frequency modulation and other relevant topics. The volume by Robinson (1974) mentioned in earlier chapters covers much of the theory of mixing and detection, especially with reference to mixing of noise with noise – though much of the original work was due to Rice (1944, 1945). King (1966) provides a useful introduction to many of the relevant noise-related topics. *Wireless World* is the source of a good many interesting and instructive articles on matters related to radio and detection: see for example Macario (1968).

18.9 Problems

1. Devise a synchrodyne receiver which uses a p.l.l. for receiving f.m. signals. Show that it is much less complex than the corresponding circuit for receiving a.m. signals.
2. Show that the conversion efficiency of a single diode switching mixer is ideally $\frac{1}{2}$ for a square-wave, compared with $\frac{1}{\pi}$ for a sine-wave.
3. Show that the conversion efficiency of a 4-quadrant multiplier used as a switching mixer is unity for a square-wave, and $\frac{2}{\pi}$ for a sine-wave.

19

Advanced Topics in Signal Recovery

19.1 Introduction

This chapter studies two major topics that were bypassed in earlier chapters, yet which merit careful analysis. The first of these involves a re-appraisal of the p.s.d. in the light of the principles of matched filtering. The second involves a study of how to make the boxcar detector resistant to offset, drift and flicker noise: this will again be seen to revolve around matched filtering concepts. In addition, it should be noted that both topics involve the use of noise-whitening filters to ensure fully optimizing the SNR.

19.2 The p.s.d. as a matched filter

So far (see especially Chapter 13) we have considered the p.s.d. as a device in which signals are chopped or modulated in the time domain, and a type of synchronous full-wave rectification is performed, in order to reduce an a.c. signal to zero frequency: this is advantageous since it permits noise to be eliminated merely by applying a much more stable ultra-narrow band filter than would be possible at the original a.c. frequency f_0. In addition, flicker noise is virtually eliminated if f_0 is above the corner frequency f_c where flicker noise drops below white noise.

At this point it is worth re-examining the operation of the p.s.d. – this time in terms of frequency. Now, the square-wave reference employed in a p.s.d. has a frequency spectrum containing many harmonics: specifically, its even harmonics are missing (this is easily understood by symmetry), and its odd harmonics have frequencies $3f_0$, $5f_0$, etc., and relative amplitudes of 1/3,

1/5, etc. In addition, the p.s.d. acts as a mixer between the signal and reference waveforms, and thus any noise components around $3f_0$ beat with the reference harmonic at $3f_0$ to produce noise around 0 Hz. Similarly for any noise components around $5f_0$, and so on (see Figure 19.1). As a result we get a total noise power proportional to $1 + 1/3^2 + 1/5^2 + \ldots = \pi^2/8 \approx 1.234$. However, an input sine-wave signal only gives output from the fundamental. Thus the voltage SNR is theoretically made $\sim 10\%$ worse by use of the p.s.d. The solution to this seems obvious – remove the noise components around the harmonics of f_0, so that only the noise accompanying the fundamental signal component is received and averaged by the final low-pass filter. Thus we have to apply an extra low-pass filter that accepts f_0 and rejects frequencies higher than $\sim 2f_0$; i.e. it is necessary to apply both an initial and a final band-limiting filter. Fortunately, the specification of the initial band-limiting filter is not at all stringent.

The arguments presented above have implicitly assumed that the ideal signal is a sine-wave. However, this is by no means universally true. For example, the thermocouple case-study discussed in Chapter 13 showed that the signal waveform may be a double exponential, and indeed, variants of this can approximate to triangular waves or square-waves – the latter probably arising rather often in practical applications. Let us consider the case of square-wave signals in more detail. If a square-wave signal is to be detected by a p.s.d., it will clearly be important to make full use of all the frequency components if the optimum SNR is to be achieved. This will mean that it is *not* a useful policy in that case to apply an initial band-limiting filter to eliminate noise at $3f_0$, $5f_0$, etc. (note that flicker noise is lower at these frequencies than it is at f_0, so this gives an extra reason for retaining these frequencies in this instance). Going back to the time-domain view of the p.s.d. merely confirms these expectations, since the synchronously

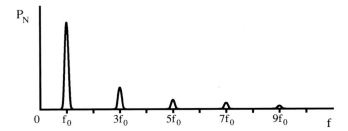

Fig. 19.1 Input noise components contributing to output noise from p.s.d. This figure shows the frequency weighting function to which the input noise power is subjected by the p.s.d. The power ratios (not drawn to scale) are 1, $1/3^2$, $1/5^2$, etc.

detected square-wave signal becomes a constant signal, and the noise, though chopped and inverted at various instants by the reference waveform, retains an identical character (careful thought shows that this is only because truly white noise contains components at all frequencies, so the noise at any point in the waveform is independent of that at any other point, however close). Overall, it appears that the p.s.d. is well suited to the detection of square-wave signals, and not (on its own) quite so well suited for the detection of other waveforms.

19.2.1 The multiplying p.s.d.

We can guess that if we wish to detect some other waveform optimally, we need to use an identical waveform as the reference. However, we have not so far allowed ourselves this option since, by definition, the p.s.d. is a device which inverts the signal whenever the reference is inverted. In fact, it is easy to generalize the p.s.d. to cover this eventuality – we merely need to employ a multiplier to combine the signal and reference waveforms. (There are some dangers in this, since (a) inversion can be implemented more accurately than multiplication, and (b) it is easy to eliminate noise on a reference waveform that oscillates between two values.) Clearly, an analogue multiplier can immediately be used as a p.s.d. for square-waves, without further change. For a sine-wave, the additional noise components at $3f_0$, $5f_0$, etc. now no longer arise – so they no longer have to be eliminated by a preliminary band-limiting filter.

So far we have dealt with sine-wave signals being detected by both types of p.s.d., but square-waves only being detected by the inverting type of p.s.d. We consider next a square-wave signal being detected by a multiplying p.s.d. with sine-wave reference. In this case, due attention will not be paid to the harmonics, and loss of signal will result. However, there is some loss of noise as well. Clearly, the loss of signal will be by a factor

$$A_{sq} = \int_0^\pi \sin \omega t \, \mathrm{sq} \, \omega t \, \mathrm{d}t \Big/ \int_0^\pi \mathrm{sq}^2 \, \omega t \, \mathrm{d}t = 2/\pi \tag{19.1}$$

where we have assumed a square-wave varying between the limits ± 1, for which:

$$\mathrm{sq} \, \omega t = \frac{4}{\pi}\left[\sin \omega t + \frac{1}{3}\sin 3\omega t + \frac{1}{5}\sin 5\omega t + \ldots \right] \tag{19.2}$$

while the loss of r.m.s. noise voltage will be by a factor:

$$B_{sq} = \frac{\pi/4}{(1^2 + 1/3^2 + 1/5^2 + \ldots)^{1/2}} = \frac{\pi/4}{\sqrt{\pi^2/8}} = \frac{1}{\sqrt{2}} \qquad (19.3)$$

Hence the overall SNR will be worsened by the same factor $\sqrt{8/\pi^2}$ (i.e. a drop of $\sim 10\%$) that arose earlier for sine-wave detection.

The multiplying p.s.d. is clearly the more powerful and general device, but the penalty for using an alternative reference waveform is (at least in the case of sine- and square-wave signals) quite small. What is more important here is that the multiplying p.s.d. is essentially a correlator, or alternatively a matched filter, supposing that it is fed with a reference identical to the expected signal, and *also* assuming that the input noise is entirely white. Of course, the latter assumption is generally invalid – as indeed we have seen since we are frequently aiming to remove flicker noise. However, the results we have obtained above appear to be valid if flicker noise or other frequency-dependent noise is *not* present. We consider below what happens if flicker noise is present. First, observe that if flicker noise is not present, the SNR is not improved by use of a p.s.d., and is actually made worse (sometimes only slightly) by use of a non-matched form of p.s.d. In fact, it is the final time-constant or low-pass filter that actually improves the SNR: the p.s.d. only makes this job easier by reducing the signal to zero frequency*.

At this point it is worth testing the earlier result for sine-wave detection, by rigorously comparing the output SNR for the two cases of inverting p.s.d. and multiplying p.s.d. The reduction in signal on using an inverting p.s.d. is:

$$A_{sin} = \int_0^\pi \mathrm{sq}\, \omega t \sin \omega t \, dt \bigg/ \int_0^\pi \sin^2 \omega t \, dt = 2/(\pi/2) = 4/\pi \qquad (19.4)$$

while the loss in noise is:

$$B_{sin} = \frac{4}{\pi}(1^2 + 1/3^2 + 1/5^2 + \ldots)^{1/2}/1 = (4/\pi)\sqrt{\pi^2/8} = \sqrt{2} \qquad (19.5)$$

---

*There is an apparent paradox here. If the p.s.d. acts as a matched filter then the SNR is automatically optimal. However, the act of multiplication does not itself make the p.s.d. a matched filter: it is necessary also to focus all the available output signal at a particular moment of time by some sort of memory circuit. In a transversal filter this is achieved by overt delay and adding circuits, but with a p.s.d. it is done in an analogue fashion by use of a time-constant. Thus it is the time-constant that finally improves the SNR, while the p.s.d. itself merely prepares the ground – though a poor p.s.d. can clearly cause a *loss* in SNR.

Hence the overall SNR is worsened by the factor $\sqrt{8/\pi^2}$ again. However, we can now deduce that the trick of placing a low-pass pre-filter before an inverting p.s.d. changes it into a *true* matched filter for a sine-wave.

19.2.2 Optimal elimination of flicker noise

Our initial purpose in using a p.s.d. was to remove flicker noise – hopefully without adding to the ubiquitous white noise. We have now seen that the latter aim is achieved. In addition, flicker noise is removed by the expedient of working at a frequency f_0 above the corner frequency f_c. But flicker noise is bound to prevent the p.s.d. from working optimally, since any matched filter requires the input noise to be white. The proper way to overcome such problems is to make the noise white by means of a suitable prefilter, observe what this filter does to the signal, and then design a special matched filter for this new type of signal. We now see that applying a noise-whitening filter *when the signal is already above the flicker noise corner frequency* has no effect on the signal spectrum, and therefore has no effect on the signal itself. Thus in this case the matched filter implementation is unchanged. However, there are more complex cases to be dealt with, such as that of the thermocouple detector case-study.

In these more complicated cases, not only does the noise depend upon frequency, but also the signal depends on frequency, and in a manner which may modify its shape as well as its amplitude. In such cases, it is no longer valid merely to increase f_0 until it exceeds f_c, but instead the SNR has to be optimized taking account of variations in *both* the signal and the noise. Unfortunately, this can only be achieved rigorously if we are prepared to redesign the matched filter for each variation in signal response – a task that would be very tedious to carry out in practice. Notice that in the thermocouple case, we optimized the SNR, but under conditions in which (a) white noise was considered small compared with flicker noise, and (b) an inverting p.s.d. was used and the possibility of optimally matching the reference to the signal waveshape was not considered. Such case studies are therefore only examples in which a scenario is worked out to see the types of situation, and solution, that can arise in practice. Overall, the most important result of this section has been the idea that the p.s.d. is a form of matched filter which will in general require a noise-whitening prefilter if the optimum SNR is to be achieved.

19.2.3 Practical p.s.d. systems

In all the above sections, ideal p.s.d.s have been assumed, whether they be inverting or multiplying devices. In particular, for the inverting p.s.d., it is assumed that the reference on and off times are exactly equal, and the normal and inverted gains are exactly equal. For the multiplying p.s.d., it is assumed that the reference waveform is exactly the ideal signal waveform (e.g. with equal duration positive and negative swings) and that the multiplying action of the device is also exact (e.g. there are no square-law terms such as might arise with an unbalanced quarter-squares multiplier). If these assumptions are not valid then the zero-frequency beat between the signal and the reference waveforms will also include, for example, offsets and noise from the reference (potentially serious, since a *small* percentage of a *large* reference voltage may be enough to swamp a small signal). In addition, notice that there may also be input noise components around $2f_0$, $4f_0$, etc, and in particular, the first of these could cause undue loss of SNR unless special measures are taken (e.g. employing a narrow-band filter on the incoming a.c. signal to ensure that any additional noise is excluded).

Modern commercially available lock-in amplifiers are capable of extracting signals at least 60 dB below noise, and are typically able to measure signals of 100 μV against a noise background of 100 mV r.m.s. Such instruments operate up to ~ 1 MHz and have input impedances ~ 100 MΩ (instruments working at higher frequencies are based on double balanced modulators and commonly have 50 Ω or 75 Ω input impedance, though these are more specialized and are not considered further here).

Important considerations in designing and operating such instruments are the *dynamic reserve* – which is the degree of overload that can be tolerated on the input signal – and the *stability* or drift per degree K in the output amplifier measured as a fraction of the full-scale output voltage. A typical lock-in may have a dynamic reserve of 1000 and a stability of $10^{-4} \mathrm{K}^{-1}$; these two figures are sometimes combined to give an overall figure of merit called the *dynamic range* – in this case $1000/10^{-4} = 10^7$ or 70 dB (the units K^{-1} are sometimes dropped, it being assumed that typical temperature drifts are ~ 1 K). However, such a figure of merit can be somewhat misleading, as there is a tradeoff between stability and overload handling capability which should be carefully observed. For signals that are subject to a great deal of interference, it is necessary to prevent system overload by setting the preamplifier for low gain, and to restore the signal to the required level by setting the output d.c. amplifier for high gain. The loss of stability (increase in drift) incurred in the latter case may well go unnoticed because of the adverse effects of the interfering signals. On the other hand, when a "clean" signal is to be detected, the input gain can be set quite high, and the output gain set correspondingly low, so that output drift is minimized.

Many modern lock-ins have various modes of operation that allow these combinations of settings to be selected at the press of a button, or in some cases an internal microprocessor can make the selection automatically.

However, such facilities do not overcome the need for careful thought to optimize signal recovery systems: for example, use of notch filters can virtually eliminate certain interfering signals, with the result that the output stability of the system may be improved by 10–20 dB. In general, preliminary signal conditioning filters can be of vital importance, and attention to such details can be especially worthwhile. However, note that frequency-domain filters can introduce significant phase shifts which might produce undesirable distortions or errors with certain signals.

19.3   Eliminating offset, drift and flicker noise with a boxcar detector

In principle, offset (i.e. the effect of a constant but unknown additive voltage) is easily eliminated with a boxcar detector by sampling both during and in the absence of a signal pulse. However, in the presence of drift (the usual situation), at least three samples are needed, of which one will sample the signal and the other two will obtain samples on either side of the signal pulse. Since we normally have to average noise, the three samples will have to be integrals rather than instantaneous values, and the overall sampling function will have a general weighting factor w; w will include a central portion that is adjusted to have the same profile as the signal to give a matched response, and we now study what the remaining contributions to w are.

Modelling the offset as a and the drift as bt, we find the apparent signal from these sources as:

$$S_{\text{apparent}} = a \int w \, dt + b \int wt \, dt \qquad (19.6)$$

and this is only reduced to zero in the general case if:

$$\int w \, dt = 0 \qquad (19.7)$$

and

$$\int wt \, dt = 0 \qquad (19.8)$$

Thus we must have a weighting function with zero mean and with a first moment that is also zero. The result in the case of a symmetric signal pulse is depicted in Figure 19.2. Notice that this is only an example, and in fact there is a further degree of freedom since the two baseline sampling regions can be shifted laterally by equal amounts in opposite directions – as in Figure 19.2(b). We can use this degree of freedom to minimize the effects of flicker noise, as flicker noise is low-frequency noise which can have the effect of making the baseline somewhat curved. This means that by moving the two baseline sampling regions symmetrically closer, the baseline at the signal pulse can be more accurately sampled. However, it is difficult to know quite where to stop with this argument. For example, should the two sampling regions overlap the outer reaches of the signal sampling region, and in any case, how should they be shaped?

Fortunately, we do not need to think for long about such matters, as there is a definite methodology for dealing with them. The point is that we are designing a matched filter for a case where the noise is not white. Hence we should start by whitening the noise with a specially designed filter, then see what this filter does to the signal profile, then design a new matched filter for optimally detecting it. To start with, the noise power spectrum is that of flicker noise (for simplicity we here ignore any white noise) which varies as

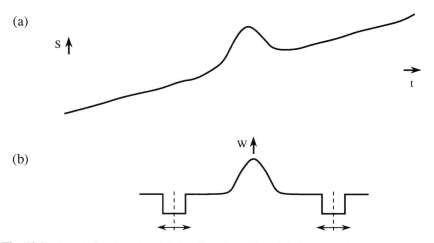

Fig. 19.2 Input signal and weighting function. Here (a) shows an input signal that is subject to noise, offset and drift; (b) shows a weighting function that can be used to suppress offset and drift. Although the magnitudes of the negative-going parts of the weighting function are easily derived, their exact positions (see arrows) are more complex to calculate (see text).

$1/f$. Thus the noise-whitening filter will have power frequency response proportional to f, while its voltage frequency response will be proportional to $f^{1/2}$; after applying this filter, the signal will be distorted, and following the theory of Section 15.5.1, we deduce that it will have to be subjected to an additional matched filter weighting factor of $f^{1/2}$. The overall signal weighting factor will therefore be $(f^{1/2})^2 = f$. Applying this in the frequency domain to the transformed signal pulse profile (i.e. the Fourier transform of the Gaussian pulse, which is itself a Gaussian) we obtain the modified frequency profile shown in Figure 19.3. Transforming this back to the time domain now gives the modified Gaussian signal pulse which is to be used as the new matched filter profile. Notice that the original Gaussian shape has become completely integrated with the offset, drift and flicker noise cancellation regions, and it is quite apparent that no amount of general discussion could have arrived at the rigorous solution we have now produced. Finally, notice that the amplitude of the Fourier transform is zero at $f = 0$, so the weighting function must have a zero mean: this would not have been so if the effects of any white noise had been included in the analysis.

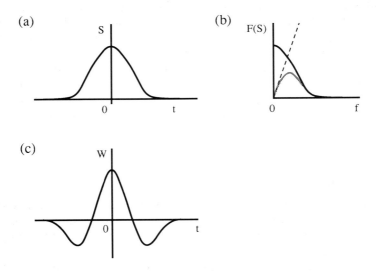

Fig. 19.3 Derivation of optimal weighting function. Here (a) shows the idealized form of the input signal, and (b) shows its Fourier transform in the frequency domain. Part (b) also shows the transform being modified by a noise-whitening filter for $1/f$ noise: when the resulting (grey) shape is transformed back to the time domain, the optimal form of the weighting function is obtained (c).

19.4 Summary

This chapter has studied two important advanced topics in signal recovery. The first is the nature of the p.s.d. – which turns out to be a form of matched filter, whose optimum form is an analogue multiplier coupled to an averaging circuit (in general the analogue multiplier has to be connected to a reference of the same shape as the expected signal waveform).

The second topic is the boxcar detector and how it reacts to offset, drift and flicker noise. In that case, too, the device in question was shown to be ideally a matched filter, in which a modified form of the expected signal is used as a reference to weight the signal being detected. Although intuitive arguments will yield an approximation to the required weighting function, an approach based on applying a noise-whitening filter followed by a suitable matched filter was found to give a rigorous solution to the problem. Thus the theory given in Section 15.5 was found to be of particular relevance.

19.5 Bibliography

The idea of treating the p.s.d. in the frequency domain is perhaps an obvious one and has been treated in numerous papers and articles since the 1950s (see for example Beers, 1965; Danby, 1970). Wilmshurst (1985) covers many other techniques for suppressing various sorts of noise and drift that had long been in wide use, but the book is creditable in systematically collecting together ideas and concepts from many disparate sources* (albeit sometimes with a tendency to hide inherently simple ideas in a mass of theory!). Unfortunately, this chapter has only been able to mention a fraction of the available techniques, but it is a subset that the author has found to be of particular generality and importance (see also Chapter 21).

19.6 Problems

1. Discuss the validity of the following principle: *anything that is known about a signal, such as its amplitude, frequency or duration, can be used to improve the SNR and hence aid signal recovery*. Give examples in support of your arguments.
2. Show that, to first order, quadrature noise (i.e. noise of phase $\frac{\pi}{2}$ relative to the signal) gives zero output power with a p.s.d., thereby ensuring an additional gain in (power) SNR of 2 for a p.s.d.

---

* In fact, the book does not list its sources, and disappointingly does not provide any references.

20

Signal Recovery and Image Processing

20.1 Introduction

This chapter aims to provide a view of signal recovery from a quite different perspective – that of the suppression of noise in digital images. The topic is also relevant in the context of applications already covered in this book, since images appear naturally as the final representation of the input data in both radar and magnetic resonance imaging. At that stage the images are being analysed by human operators and clarity of presentation is important, so noise suppression is vital – as indeed it is in other situations where images are presented to humans. Here, images from space probes and guided missile sensors, industrial and security cameras, text readers, thermal imagers and medical equipment of all types come quickly to mind.

20.2 Image noise suppression filters

We have already seen that median filters can be applied to sampled waveforms in order to remove noise spikes. Here we consider how such filters are applied to image data. Figure 20.1(a) shows an image that has been digitized into a large number of picture cells or *pixels*. Some noise in the various pixel

Fig. 20.1 Examples of filtered images. Here (a) shows a grey-scale image digitized into a 128×128 array of pixels. Parts (b), (c) and (d) show respectively the results of filtering the image in a 3×3 processing window using a mean filter, a median filter, and a mode filter. Notice that in all cases, noise is successfully suppressed, but in (b) the edges of the objects are blurred, and in (d) they are enhanced. Part (c) gives by far the most faithful reproduction of the underlying signal.

Fig. 20.1(a)

Fig. 20.1(b)

Fig. 20.1(c)

Fig. 20.1(d)

intensity values is apparent, though the eye itself tends to ignore such noise because of the intelligence it applies to interpreting any image, so it is easy to underestimate the amount of noise that is actually present. However, it is clearly useful to apply noise suppression filters to such images.

The median filter is probably the most widely used type of filter for digital images. The reason for this is that a normal mean (moving-average) filter, with a 2-D convolution kernel of the form:

$$M_1 = \frac{1}{9}\begin{bmatrix} 1 & 1 & 1 \\ 1 & 1 & 1 \\ 1 & 1 & 1 \end{bmatrix} \quad \text{or} \quad M_2 = \frac{1}{16}\begin{bmatrix} 1 & 2 & 1 \\ 2 & 4 & 2 \\ 1 & 2 & 1 \end{bmatrix} \quad (20.1)$$

tends to blur the underlying signal as shown in Figure 20.1(b), as well as removing noise, so some form of filter that avoids this is required. Now M_1 and M_2 can also be regarded as approximations to low-pass frequency-domain filters, so the latter type of filter would not be any different in its blurring qualities. Thus the median filter is valuable because of its capability for removing noise without introducing blurring (see Figure 20.1(c)).

But what gives the median filter this capability? The answer lies in its makeup, i.e. in its particular way of calculating the mean. First, for the region around each pixel it obtains a histogram which contains the distribution of local intensity values. Then it eliminates the outliers systematically, by cutting out the extreme values at each end of the distribution, and repeating the process until only the middle (median) value remains*. It then takes this as the new optimum local intensity value, building up a new image from all such values. Clearly, for a monotonically increasing (or decreasing) function, the result of averaging with this procedure leaves the window centre value unchanged, so no blurring (mixing in of distant values) can occur.

It turns out that the median filter is excellent at removing impulse noise. The reasons for this can be seen from the 1-D case shown in Figures 14.5 and 14.6 (see Section 14.5). Basically, this is because any impulse is not averaged into the output data, but is totally ignored: the algorithm thus acts in an "intelligent" way. The possibility of its being more intelligent lies in its highly non-linear mode of operation (see the detailed algorithm in Table 20.1). On the other hand it turns out that the median filter is slightly less effective at removing Gaussian noise than the mean (moving-average) type of filter. Thus some care must be taken to determine exactly what types of noise images are subject to, since there is a possibility that the mean filter

---

* Statistically, the median is defined as the value which divides the area of the distribution into two equal parts.

Table 20.1 Implementation of median filter algorithm.

```
begin {run over image}
   [[ for i:=0 to 255 do hist[i]:=0; {clear local intensity histogram}
      for m:=0 to 8 do hist[ P[m] ]:=hist[ P[m] ]+1;
                                      {enter intensities in histogram}

   i:=0; area:= 0; {set pointer i at left of histogram}
   while area <5 do {run towards right of histogram}
     begin
        area:=area+hist[i]; {find area to current point in histogram}
        i:=i+1; {move to next position right in histogram}
     end;
   Q[0]:=i−1 {output intensity value of previous position in histogram} ]];
     end;
```

In this algorithm, the median intensity value is the 5th out of 9 in a 3×3 processing window, so the pointer has to move towards the right of the histogram until the 5th intensity value is passed, and then backtrack by one.

Notation: the original image is placed in *P*-space, and the processed image appears in *Q*-space. The labelling of pixels in a 3×3 processing window at a general location in (for example) *P*-space is:

P[4]	P[3]	P[2]
P[5]	P[0]	P[1]
P[6]	P[7]	P[8]

The double square brackets [[...]] indicate the double for-loop required to scan over a 2-D image. In this notation, the algorithm for a 2-D moving-average mean filter (M_1 in equation (20.1)) would be:

```
begin {run over image}
   [[   sum:=0;
        for m:=0 to 8 do sum:=sum+P[m];
        {add intensities of pixels in window}
        Q[0]:=sum/9 {output average intensity value} ]];
   end;
```

will give better results. Factors to take into account are the signal bandwidth, the noise bandwidth, and the widths and heights of any impulse noise spikes. However, we shall not pursue this matter further here, but merely leave the hint that the principles that guide the suppression of noise in images might also be valid in more conventional electronic systems where signal recovery is being performed.

It is also of interest to note that there are many other types of filter. In particular, we have already covered mean and median filtering, and this raises the question of whether mode filtering might be useful. After all, the mode of any distribution is the most probable value of that distribution – immediately suggesting that it should give improvements over both of these other types of filter. (Of course, in a symmetric distribution, all three means are coincident, and then it will not matter which of them is employed: however, when for example a spike noise impulse is present, the local intensity distribution will be skewed, so in that case it will definitely matter which of the three means is taken.)

Experiments with mode filters quickly show that they remove a certain amount of noise (this is so since they eliminate intensity values at each end of the local intensity distribution), though they also have another effect – that of enhancing edges in an image (see Figure 20.1(d)). This latter property arises since taking the mode of the distribution means that any secondary mode corresponding to a minority of intensities near the edge of an object will result in background being ignored if the centre pixel is within the foreground, whereas the foreground will be ignored if the centre pixel is within the background. Hence the overall effect is to concentrate on relevant local foreground intensities in the first case, and on relevant local background intensities in the second case – thereby sharpening up the intensity profile as shown in Figure 20.2. Again, this is an "intelligent" type of processing,

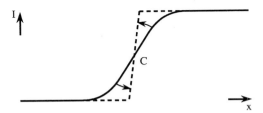

Fig. 20.2 Edge enhancement by the mode filter. Here, the original edge profile (solid line) is modified by the action of the mode filter to that shown by the dotted line. If the centre of the processing window is to the left of the profile centre C, the mode lies approximately at the minimum of intensity and the higher intensity values are ignored: the opposite happens to the right of C.

and far from resulting in a blurred image, it tends to de-blur images. We shall not pursue this topic further, but note that it implies a lot about interaction between recognition and noise suppression, both of which underlie the concept of signal recovery.

20.2.1   Shifts of curved edges

We have seen above that noise suppression without blurring can be achieved by at least two types of filter. Returning to consider in more detail the action of the median filter, we find next that it has one inconvenient property – that it shifts certain types of edge in an image. By symmetry, there can be no such shift for a straight edge with an S-symmetric edge intensity profile. On the other hand, for curved edges this does not apply, and in fact there are small shifts *inwards* towards the local centres of curvature in such cases. This means that objects become slightly smaller when processed by a median filter! In fact, very small objects are eliminated by a median filter (this has to be so as it is the basis on which spike noise is removed). Theory shows that the shift is:

$$S \approx \frac{a^2}{6\rho} \qquad (20.2)$$

where ρ is the local radius of curvature, and a is the radius of the median filtering window. Experiment shows that this formula is not exact, and the true situation is indicated in the graph of Figure 20.3. This graph can be used to correct any measurements that are taken from an image which has been processed by a median filter, this being an example of intelligent high-level processing being applied to remove deficiencies of relatively *ad hoc* low-level algorithms. There are certain other filters that do not suffer from such distortions to quite the same extent. However, they do still give distortions, and by using them we do not avoid the inherent problem. Thus calibration curves of the type shown in Figure 20.3 may still be needed. Finally, we do not escape from these distorting effects by reverting to use of a mean filter – almost identical shifts and distortions have been found to arise in this case too, and equation (20.2) still applies, albeit with slightly modified numerical coefficients (depending on the type of edge profile).

20.2.2   Corner detection

Next, we briefly consider applying the median filter's distorting properties to useful effect. In particular, if objects are approximately polygonal, and

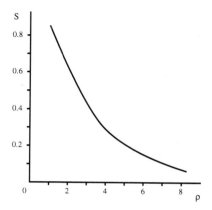

Fig. 20.3 Shift S of edge position as a function of curvature. Here, S is the local shift in edge position that is caused by a median filter at a point where the local radius of curvature is ρ. Both scales are in pixels. This curve resulted from experiments (Davies, 1989) using a median filter with a 5×5 processing window.

have fairly sharp corners, the median filter will have the effect of cutting off the corners while leaving the remainder of each shape nearly unchanged. Hence subtracting the new shape from the original shape will give a "corner picture" of any object! This effect forms the basis for a corner detector that can be employed for object location (though there are clearly issues about how corners can be discriminated from noise by this means – space will not permit a full discussion of this here).

20.2.3   Discussion

This book is not especially concerned with image processing and image analysis, and therefore discussion of further topics relating to object location (or "recovery") in images is deferred. Suffice it to say that the principles of signal recovery and matched filtering discussed in earlier chapters are relevant to image analysis, though they have to be adapted (a) for the specific case of 2-D, (b) for the much greater amount of data to be handled in this case, and (c) to allow for the fact that in this application area the processing will have to be carried out by digital computers, sometimes with the help of dedicated fast digital hardware. Clearly, image analysis is a test case showing the way conventional signal recovery has to be adapted to modern processing methods.

20.3   Summary

This chapter has discussed noise suppression as a form of signal recovery in digital images. The median filter is especially important in this context, since it is able to suppress noise without blurring the image signals – though it is found to shift edges slightly towards local centres of curvature. This effect is not avoided with mean filters, though it can be *reduced* using certain types of "detail-preserving" filter. However, the error curve is known and can be used to correct any measurements made from a median-filtered image. The distorted image can also be used for locating corners of polygonal objects, to aid recognition. Finally, the mode filter is useful both for removing noise and for enhancing object boundaries: thus both the median and the mode filter are seen to embody a certain intelligence, though this must not be over-stated as *really* intelligent processing only arises when image features are passed on to higher-level algorithms for much more abstract kinds of processing.

20.4   Bibliography

All the topics discussed in this chapter are covered in more depth in the author's book on machine vision (Davies, 1990) – see especially Chapter 3. Davies (1991) provides a proof that the mean filter also suffers from inward shifts towards the local centres of curvature; Davies (1989, 1992a,b) makes other relevant comparisons. Matched filters are mainly used in image analysis/machine vision for detecting low-level features such as the edges or corners of objects (Davies, 1992c).

Another form the matched filter often takes in machine vision is best described as in the last few lines of Section 20.3, namely that image features are passed to higher-level algorithms: specifically, the Hough transform is frequently used for this purpose (see Davies, 1990, Chapters 8–15). Discussion is curtailed in the present volume because such topics are rather specialized and roam well outside the purview of conventional signal recovery.

20.5   Problems

1. Show that a median filter acting within a 3×3 square window will cut off the corner pixel in the following image section:

0	0	0	0	0
0	0	0	0	0
0	0	1	1	1
0	0	1	1	1
0	0	1	1	1

2. Show that, because of the neighbouring noise spike, a median filter acting within a 3×3 square window will shift the edge in the following image section, though the spike itself will be eliminated:

0	1	0	5	4
1	0	0	6	5
0	5	0	5	5
0	6	0	5	6
0	0	1	5	5
1	0	0	4	4

3. If a circular object of radius ρ is just eliminated by a median filter operating in an ideal circular window of radius a, determine the effective edge shift S that occurs. Show that this result is not accurately represented by equation (20.2), though it is compatible with the graph of Figure 20.3.

21

Putting It All In Perspective

21.1 Introduction

This chapter aims to provide some perspective on the material covered in this book, and also to make some general comments about the nature of signal recovery. To some extent space has been a limiting factor, preventing related topics such as signal processing from being dealt with in depth; some justification will therefore be given for the particular stance taken. We start in Section 21.2 with an appraisal of the matched filtering technique.

21.2 An appraisal of matched filtering

Part 1 of this book described various circuits for amplifiers, oscillators, power supplies and so on, and perhaps gave the impression that electronics is purely about producing a specification and then rather mechanically designing a circuit to fulfil some necessary task. Part 2 brought things down to earth by showing that noise is an awkward fact of life that must be minimized – e.g. by careful adjustment of circuit impedances. However, Part 3 brought us to certain applications such as radio, radar, and magnetic resonance where we frequently have to fight rather hard to recover signals from below the noise barrier.

In the end, we achieved considerable success in this endeavour, and having emerged from the dark tunnel, it is useful to analyse how this progress was attained. In fact, there were several independent means of reaching these goals. These involved use of a barrage of devices: frequency-domain filters, p.s.d.s, boxcar detectors, signal averagers, noise-whitening filters, matched filters, superhets and phase-locked loops. However, closer examination reveals that several of these devices are related to each other. For example, synchronous detection, which uses phase-locked loops, is really the same as phase-sensitive detection, and the phase-locked loop itself contains a p.s.d. Noise-whitening filters are normally forms of frequency-domain filter, and p.s.d.s are forms of ultra-narrow band frequency-domain filter. In addition, a.c. detection is valuable as a means of preferentially eliminating flicker noise, and again the superhet performs this feat by converting h.f. input waveforms not to d.c., but to some suitable i.f.

At this stage, we still have a number of seemingly distinct approaches. However, in Chapter 19 we went further with our analysis of the action of p.s.d.s and boxcar detectors and found that they can be considered as special cases of the matched filter. Indeed, this interpretation proved valuable, since we could then apply noise-whitening filters to fully optimize the final SNR.

At this point there appears to be just one general approach to signal recovery – that of matched filtering. The question now is whether matched filtering is the only really fundamental means of enhancing SNR.

To answer this question, let us re-examine our list of devices for signal recovery. Consider signal averaging. We have already seen in Chapter 19 that one form of signal averaging is identical to matched filtering – that is the use of a cleverly weighted filter profile in order to eliminate offset, drift and flicker noise. However, we are still left with another form – which is used in the boxcar detector: in this case many identical signals are added so that although both the signals and the noise powers accumulate, the SNR is also augmented. We can now see that the matched filter concept can be generalized yet again, to add many separate versions of a signal, by considering *the set of separate instances of the signal as constituting one global signal whose SNR is to be enhanced* (i.e. the filter has to be matched to a waveform consisting of a repeated signal – see Figure 21.1). Thus our thesis about the power of the matched filter has expanded again.

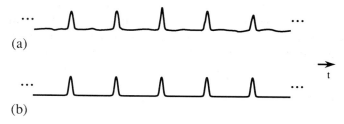

Fig. 21.1   Matched filter for a multiple bleep signal. This figure shows (a) a signal consisting of a sequence of many "bleeps" (such as radar echoes) which are to be averaged, and (b) the global matched filter that has to be applied in this case.

21.2.1   Limitations

What then are the limitations of this concept? There appear to be two (apart from the obvious one that matched filters can only be used when we have prior knowledge of the signal profile). The first is that signals do not just appear: they need to be *stimulated* to appear, and when they appear they will do so in one specific form resulting from the means by which they have been stimulated. Clearly, we are able to code signals in various ways, and in particular we can code them so that they avoid excessive contamination by known types of noise. Thus our original a.c. detection approach was devised specifically to avoid flicker noise: the signals were coded to make them more readily distinguishable from this type of noise. Matched filtering only gave the *actual means* for distinguishing them.

The second limitation concerns the types of noise that can arise. Now matched filtering is only provably optimal when noise is white, and in other cases noise-whitening filters have to be applied. Yet there are cases when such filters are difficult or even impossible to design. One such case is in image analysis of everyday scenes, where the background is far from pure white noise, but consists of a jumble of objects of various sizes, shapes, colours and textures: there can only be approximations to noise-whitening filters in such cases. Another such case is that of impulse noise, which is poorly discriminated by linear filters – yet non-linear filters such as median filters are able to eliminate this sort of noise with relative ease. These facts give the impression that matched filters are not at all general.

21.2.2   Summary

Thus our view at this stage is that when complex systems are stimulated to give particular types of information, such a plethora of signals and noise will emerge that recognizing the desired information will be difficult and the associated SNR will be low – *unless* we can design both the stimulation and the final filter together: thus it will not be optimal to design the stimulation system and then the matched filter, but there must be some methodology for arriving at a globally optimal system. However, it appears that no such methodology is currently available.

Let us next backtrack a little to see if we can at least tackle the other main limitation – that of the desired signal appearing in a background of irrelevant signals rather than pure noise. Clearly, we have to use a highly non-linear filter for this purpose. Notice, however, that the human eye and ear regularly perform prodigious feats of recognition which it would be hard for a machine to emulate – locking onto a conversation at a party, recognizing (acoustic) footsteps in a corridor, and so on. It is intelligence (coupled with a large database) that permits such feats to be performed. Thus the proper way to proceed with signal recovery must be to move to high-level signal processing with large rule-based systems: our matched filter approach is surely limited here. On the other hand it has, as we have seen earlier in this book, a quite impressive record which we should not eschew or belittle. Having a thoroughly understood mathematically-optimal stepping stone is a good place to start for searching new worlds of recognition and identification, whether we can code signals as we wish or have arbitrary codes thrust upon us. This book has not attempted to make inroads into these more advanced areas: an extra chapter on artificial intelligence would not have achieved very much*, but the path seems sufficiently mapped out to leave the problem an open one for the reader to ponder about.

21.3   Digital v. analogue electronics

The knowledgeable reader will probably feel that this volume has been somewhat reticent to discuss digital electronics. There are two reasons for this. First, in reasonable space, it is difficult for a book to cover in depth both digital and analogue electronics, as well as signal recovery methods and applications; and of course there is always a fourth topic – digital computation and microprocessor electronics – waiting in the wings for its turn.

Second, it turns out that many of the principles of signal recovery are analogue in nature, digital electronics and digital computation being exceedingly valuable for various reasons concerned with implementation, but not on the whole leading to significant new possibilities for signal recovery†. We now explore some of these points. But first note that time-domain switching is here being taken as an analogue technique.

Basic analogue concepts and principles that are relevant to signal recovery are:

1. Frequency-domain filtering, as performed in many types of low-pass, high-pass, band-pass and band-reject filters, and noise-whitening filters.
2. Time-domain switching, as used in the p.s.d. and boxcar detector.

---

* Many (whole) books are already available on such topics: see for example Winston (1984).
† Whilst this is true for conventional forms of "pure" signal recovery, as defined in this book, it is certainly invalid for recovery (in the sense of *recognition*) of complex signals in a general background of clutter, as discussed in the previous section: see also later parts of the current section.

3. Signal weighting and averaging, as performed by signal averagers, matched
   filters, time-constants, etc.

This list covers most of the methods described in this book. Probably, it
is best to consider what digital electronics would do *better*. On the whole,
digital electronics is considerably more accurate and free from drift, though
it is questionable whether it processes signals any faster. On the other hand,
if we include digital computers, these allow large databases to be built up
on the data being processed, and hence aid the application of intelligent
processing – a vital factor in the development of advanced processing
methods. Furthermore, complex algorithms which cannot be implemented
quickly in raw analogue or digital electronics can rapidly be tried out on a
digital computer, so digital computers really do permit much more to be
achieved.

An important example of the use of a digital computer is the processing
of images and acoustic data to eliminate impulse noise or complex clutter
signals. Median filters provide a simple example: although the algorithms
for median filtering are more complex than those of moving-average filtering,
they hardly bring a computer to its knees (except with enormous images!).

Finally, the sampling theorem provides the basis for assuming an equivalence between analogue and digital techniques and thereby stresses the
importance of considering basic principles separately from implementation.

Overall, this book has found it possible to discuss most of the relevant
issues in signal recovery without delving into complex algorithm design.
However, the previous section has made it clear that this is indeed the next
stage, and practitioners who are keen to separate electronics into mutually
exclusive headings – analogue, digital, computation, and so on – should
beware: signal recovery is a meeting ground for techniques in all these
categories.

21.4 Summary

This final chapter has taken the idea of matched filtering further and has
demonstrated that it includes several other important methods of signal
recovery as special cases: in particular, phase-sensitive detection, boxcar
detection and signal averaging fall into this category. This conclusion focuses
attention on what processes do *not* fall within the matched filtering formalism.
One such process is coding of signals in such a way that they are easier to
identify; another is the use of non-linear filtering (especially median filtering)
to eliminate impulse noise and outliers; and a third is the use of "intelligent"

processing to eliminate clutter signals arising from currently irrelevant objects. Overall, the fact that matched filters have the limitations indicated above is less remarkable than the amount these rather simple linear processors *are* able to achieve.

21.5 Bibliography

It is difficult to give a bibliography for the topics covered in this chapter, or for the smaller area of signal recovery. Generally relevant books include those by Robinson (1974), Wilmshurst (1985) and Horowitz and Hill (1989), while the author's book on machine vision (Davies, 1990) covers certain topics that are relevant for situations where complex signals have to be extracted from backgrounds containing considerable clutter. For clear accounts of the subject of artificial intelligence, see for example Winston (1984), Charniak and McDermott (1985). For further specific references, the reader is referred to the bibliography sections of previous chapters.

21.6 Problems

1. List the advantages of computers for signal recovery and signal processing.
2. List the processes underlying signal recovery, stating clearly which are analogue and which are digital in nature.
3. A 1-D square-wave signal is to be detected using a matched filter. Determine the shape of the response after passing through the filter: show that the temporal resolution is reduced if the impulse response is set slightly too short *or* slightly too long. Show also that this result is peculiar to a square-wave signal, and is not replicated if, for example, a Gaussian signal is to be detected by an appropriate matched filter.

Appendix A

Semiconductors

An *intrinsic* (or *i-type*) semiconductor is a pure semiconducting material (such as germanium or silicon) in which the valence band is full and the conduction band is empty at the absolute zero of temperature. At normal temperatures a number of thermally generated electron–hole pairs exist (some electrons from the valence band have been excited to the conduction band, leaving *holes* in the valence band), and these carriers are able to conduct weakly in an applied electric field. Very small currents are generally of little use in electronic devices, and so special impurities are added at low concentrations to the pure semiconducting material, in order to provide extra charge carriers. Thus boron impurities (valence 3) act as *acceptors* which create additional holes, while phosphorus impurities (valence 5) act as *donors* providing additional electrons.

A semiconducting material with predominantly donor atoms is called *n-type*, and one with predominantly acceptor atoms is called *p-type*, because of the nature of the charge carriers that are present in either case. When an electron appears in an n-type semiconducting material it is regarded as a *majority carrier*, and similarly for holes in p-type material; when electrons appear in p-type semiconducting material, or holes in n-type material, they are regarded as *minority carriers*.

In the FET, majority carriers carry the current through the main conducting channel. However, in the BJT, majority carriers carry the current in the emitter region, but in the base region these carriers become minority carriers, while in the collector region they again become majority carriers. On the other hand, majority carriers in the base region are important in providing the base current and in annihilating a proportion of the minority carriers passing through the region. Since the properties of the BJT are primarily determined by the fate of the minority carriers in the base region, the BJT is sometimes called a *minority carrier device*, whereas the FET is called a *majority carrier device*. Likewise, in the BJT both types of carrier are important in determining the properties of the device, so it is also termed a *bipolar device* (hence the name *bipolar junction transistor*, BJT), whilst the FET is termed a *unipolar device*.

Finally, the BJT is often considered to be a *current operated device*, since the ratio collector current/base current is large and approximately constant, and the large current gain gives the device its amplifying properties. By contrast, the FET is considered to be a *voltage operated device* (hence the name *field-effect transistor*, FET), since scarcely any gate current flows (except in the form of transients which charge the capacitance of the gate junction), and it is the gate voltage that controls the current flowing through the channel, and which eventually results in the FET's amplifying properties.

A.1 Bibliography

For further information on solid state physics, the reader is referred to Kittel (1966). Sze (1981) deals specifically with solid state devices. A more elementary treatment of the subject may be found in, for example, Calvert and McCausland (1978).

Appendix B

Electronic Devices

A *varactor* (variable capacitance) diode is a p-n junction that has a significant capacitance when it is reverse-biased. A variable reverse bias voltage V is able to vary the separation of the majority carriers in the two regions, and hence changes the capacitance of the device as seen by a small a.c. voltage v. Thus the device can be used for remote control of radio tuners, or for instituting automatic frequency control around a chosen station. It is also used in certain types of parametric amplifier (in which a "pump" is applied at one frequency, to feed in power, allowing the device to amplify at another frequency).

A *tunnel diode* (or *Esaki diode*) is a diode with an especially narrow junction region, through which quantum mechanical tunnelling can occur. This gives the device a region of negative resistance over a small range of voltage. As a result it can be used in oscillators, h.f. amplifiers and fast switches: further details are beyond the scope of the present text.

A *back diode* is a form of tunnel diode which is operated below the negative resistance region where the current–voltage characteristic has a sharp knee, giving the device a very low switch-on voltage (~ 0.1 V): this permits it to act as an efficient detector of low-level a.c. signals.

The *thyristor or silicon controlled rectifier* (SCR) is a 3-terminal device that is triggered by a voltage pulse on its gate, and then remains in a conducting state until the current between anode and cathode is brought down to zero. It finds considerable use for providing over-voltage protection in power supplies, and also for switching a.c. waveforms – e.g. to control and dim electric lights. (See also Problem 9.4.)

The *Schottky* (*hot carrier*) *diode* is a metal–semiconductor junction that has a rather low switch-on voltage (~ 0.2 V), which makes it suitable both as a detector of low-level a.c. signals and as a means, when placed across the collector-base junction of a BJT, of preventing saturation, thereby permitting operation of the BJT as a high-speed switch.

The *thermionic valve* is a device in which electrons travel from a heated cathode through a vacuum to an anode collector, held at a high positive voltage. The *thermionic diode* (which contains just a cathode and an anode) can be used as a rectifier. The *thermionic triode* also has a grid of wires between the cathode and the anode; small voltage changes on the grid can produce considerable changes in the anode current, giving a high mutual conductance, and thereby allowing the device to act as a useful amplifier. Developments of the triode include the *tetrode* and the *pentode*, which achieve higher gain. However, the advent of semiconductor devices in the 1950s spelled death to most types of thermionic valve. The survivors are the magnetron (a highly specialist device used for generating high-power microwave pulses), and very high-power devices used in the output stages of radio transmitters: semiconductor devices are still unable to compete with valves when powers in the kilowatt to megawatt range are involved.

For the p-n junction diode, bipolar junction transistor (BJT), and field-effect transistor (FET) see Chapter 1 and Appendix A.

A *choke* is an inductor that is used to block the path of high frequencies. For example, in radio receivers *radio-frequency chokes* (r.f.c.s) are often used in conjunction with *feed-through capacitors* to prevent r.f. from passing along the power-supply rails.

A *surface-acoustic wave* (SAW) device is a block of solid piezo-electric material in which a sound wave is launched by applying oscillating voltages to metal electrodes on the surface. To achieve high efficiency over a narrow band of frequencies, a periodic "interdigital" set of electrodes is employed to launch the wave. In addition, an interdigital set of electrodes is employed to pick up the wave at a later time. These devices are valuable for introducing long, stable time-delays of up to $\sim 50\,\mu s$ in electronic circuits. As a result they find application in frequency-domain filters and in matched filters for radar. To achieve longer delays of up to ~ 1 ms, devices can be cascaded or other special techniques may be used: details are beyond the scope of this book, though the ideas are intriguing and ingenious and merit careful study.

The *charge-coupled device* (CCD) is a form of analogue shift register in which packets of charge are passed from one location to another in a long sequence, as clock pulses are fed in. With a 3-phase clock, charge is fed from location $3i$ to location $3i+1$ ($i=1$ to n), then from location $3i+1$ to location $3i+2$ ($i=1$ to n), then from location $3i+2$ to location $3i+3$ ($i=1$ to n). Most CCDs now operate with a 2-phase clock, but this requires asymmetric cell design. Losses of charge do occur, but by now linear CCDs of 4096 elements are practicable, and whole (2-D) images of size 512×512 pixels or larger can be obtained from CCD cameras. Indeed, most modern video cameras are fabricated using CCD technology. Finally, the controlled time-delays of CCDs permit use as transversal filters and in particular as matched filters.

Analogue-to-digital and *digital-to-analogue converters* (ADCs and DACs) are nowadays vital for converting analogue waveforms into digital format, and finally (if necessary) back to analogue format, so that digital computers and dedicated digital hardware can be used for precision processing of data. One of the main restrictions on their use is speed of operation. However, 8-bit "flash" ADC chips are available which operate at rates of over 200 megasamples per second, while DACs with similar specifications are if anything easier to design. Details are beyond the scope of the present volume.

B.1   Bibliography

Most of the devices mentioned in this appendix are described in reasonable detail in Horowitz and Hill (1989). SAW devices are more specialized: a thorough, fairly up-to-date account is provided by Morgan (1985), with special reference to signal processing applications; see also the early popular account in *Scientific American* by Kino and Shaw (1972). For CCD technology, see the volume by Beynon and Lamb (1980), and the early article by Amelio (1974). Digital circuits such as shift registers, and also ADCs and DACs, are covered in very many books on digital electronics, such as Strong (1991); see also Horowitz and Hill (1989).

Appendix C

Microwave Devices

This appendix is intended to cover, at a basic level, the microwave devices and techniques needed for an understanding of the radar and EPR systems described in Chapters 13–17. It is also relevant for the topics of radio astronomy and satellite communications which were touched on in Chapter 11. It covers in turn generation and amplification, transmission, processing, coupling and detection of microwave radiation.

C.1 Sources and amplifiers of microwave radiation

Perhaps the most important source of microwave radiation is the *cavity magnetron*, so-called because it employs a series of resonant cavities and a strong magnetic field. Electrons emitted by a cathode are caused to circulate around it by the magnetic field and then energize the cavities. This type of device is capable of generating several megawatts of microwave power, and is normally used in pulsed operation, especially for light airborne radar transmitters – though it has recently come into prominence for microwave ovens! Particular advantages of the device are its relative compactness and ruggedness.

The *klystron*, by contrast, is basically a microwave amplifier capable of producing up to $\sim 20\,\mathrm{MW}$ of power in pulsed mode; it works by causing "bunching" of electrons travelling from one input cavity resonator to another where the amplified microwave power is extracted. However, in a *reflex klystron*, the beam of electrons is reflected from an electrode at a negative potential, and positive feedback via a single cavity resonator takes place. Reflex klystrons are capable of working continuously and produce up to

~ 1 watt of power: thus they are suitable for use as local oscillators in microwave superhet receivers.

The *Gunn diode* is a GaAs diode which is mounted in a small microwave cavity and oscillates at microwave frequencies, producing up to ~ 250 mW of power. This makes it highly convenient for use as a local oscillator for microwave receivers. The *IMPATT diode* has a more complex internal structure than the Gunn diode and is capable of providing slightly more microwave power, while the related *TRAPATT* devices are able to provide several hundred watts of pulsed microwave power at duty cycles of up to 1%. TRAPATT devices are therefore suitable for short-range pulse radar transmitters. Note that the Gunn diode produces less noise than the IMPATT diode, thereby enhancing its suitability as a local oscillator.

A *travelling wave tube* is a device containing an electron beam along which an electromagnetic wave can be guided so that energy is transferred from the beam to the wave. Thus the device acts as a microwave amplifier. It has the advantage that it will operate satisfactorily at high powers (up to ~ 100 MW) and in addition has a wide bandwidth. This latter property enables it to be used for amplifying radar "chirp" signals, and in "frequency agile" radars.

The *maser* is a microwave amplifier which works by the stimulated emission process (the term "maser" is an acronym for *microwave amplification by stimulated emission of radiation*). Here atoms are pumped by an r.f. source so that the relevant quantum states undergo a population inversion. Then incoming radiation at an appropriate frequency is able to stimulate excited atoms to emit at the same frequency and thereby augment the signal. The resulting gain in power is only possible because energy has been fed to the device by the pumping source. Masers have the lowest noise figure of any type of microwave amplifier, their intrinsic noise temperature T_N corresponding roughly to the temperature T at which they operate: in fact, the value of T_N depends on the ratio of the signal frequency to the pump frequency, and as a result it is possible for T_N to be *lower* than T. For example, a maser cooled with liquid helium to 4.2 K may have an intrinsic noise temperature of ~ 3 K while delivering some 30 dB of gain, though as we have seen in Section 11.4, losses can make the system noise temperature much higher.

Parametric amplifiers produce their gain by modulating some convenient circuit or device parameter, and this permits power to be fed into the system and eventually used for amplification. Perhaps the best known and most widely used type of parametric amplifier employs a variable-capacitance diode. The pump frequency modulates the capacitance, and this causes mixing with the incoming signal frequency: the output at the difference frequency then has increased power. Ideally, the noise figure of such a system depends only on the diode temperature and the ratios of the frequencies, and there are no resistive losses to affect it. Clearly, this can only be an approximation in practice – or even in principle, since the output circuit has to absorb power and must therefore be resistive, lossy and noisy. In spite of this, noise temperatures as low as 12 K can be achieved with diodes cooled to 4 K.

Both masers and parametric amplifiers are essentially narrow-band devices. The maser is appropriate for frequencies in excess of 1 GHz, while the parametric amplifier is of little advantage over conventional transistor amplifiers below 500 MHz: with the advent of the new low-noise GaAs FETs (especially HEMTs), this lower limit is gradually being pushed upwards.

C.2   Transmission of microwave radiation

Microwave radiation is a form of electromagnetic radiation, normally in the frequency range $10^9 - 10^{11}$ Hz (wavelength 30 cm down to 0.3 cm), and travels by line of sight in free space. It will also travel down transmission lines, but is subject to significant dielectric losses in coaxial cables and radiation losses in open transmission lines. For these reasons it has been customary to cut down losses by use of *waveguide* – copper or other tubing, normally of rectangular cross-section, without any centre conductor (Figure C.1): waveguide works by reflecting the microwave beam continually from side to side along its length (this action gives a resultant beam directly along the waveguide, but only when the free-space wavelength of the radiation is less than twice the width of the waveguide). For a clean waveguide with good surface conductivity, losses are normally low, and tracts of many metres are acceptable.

In compact electronic equipment, waveguide is too heavy and bulky, and nowadays *stripline* (which may be envisaged as a compacted form of coaxial cable constructed using dielectric) and *microstrip* (a compacted form of twin-pair cable in which one line is replaced by a ground plane, and the other is separated by a dielectric) is used (Figure C.2). The latter is akin to printed circuit board construction and is well-adapted for use with active and passive components. However, it is liable to radiate, and is therefore normally only used in completely enclosed equipment, such as complete local oscillator, mixer and bandpass filter assemblies.

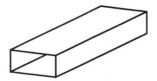

Fig. C.1   Oblique view of a section of a waveguide.

(a)

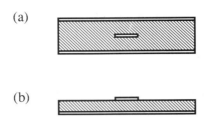

(b)

Fig. C.2 Compact forms of microwave transmission line. (a) The cross-section of a *stripline*, which can be regarded as a compacted form of coaxial cable. (b) The cross-section of *microstrip*, which can be regarded as a compacted form of twin-pair cable (one of the pair is removed and replaced by a ground plane).

C.3   Miscellaneous devices for processing microwave radiation

An *isolator* is a device which permits microwave radiation to pass down a waveguide (or stripline, etc.) in only one direction. Such devices are normally inserted in a waveguide from a microwave source in order to prevent radiation being passed back into the source and thereby affecting its operation – e.g. by "pulling" its frequency. An isolator is constructed using a special material, such as a ferrite bead biased by a permanent magnet, since most passive microwave circuits necessarily conduct equally in each direction.

A *TR-cell* (a transmit-receive cell, also known as a *duplexer*) is a device which is normally non-conducting and which does not interfere with microwaves passing along a waveguide in which it is mounted. However, a high-power pulse from a magnetron can cause the gas in it to ionize and to conduct strongly until the pulse has passed. Thus the device can protect a delicate microwave receiver connected to the same antenna as a transmitting magnetron.

A *PIN diode* (which consists of three thin layers of semiconducting material, of types p, i, n respectively) is an active device which may be made to conduct by an externally applied current, thereby causing a short circuit across a waveguide. A set of three or four of these mounted at carefully chosen intervals along a waveguide is capable of cutting down the microwave power by 40 dB or more. The device is intended for low power and high isolation, whereas a TR-cell is intended for high power, but moderate isolation. In a radar receiver, a combination of the two is generally the most appropriate.

A *matched load* is a device that is inserted at the end of an unused waveguide run in order to absorb completely any microwaves passing into it. This prevents reflections from the end from affecting microwave components further back down the waveguide.

A (variable) *attenuator* is a metal vane that may be moved from the walls towards the centre of the waveguide, thereby absorbing increasing fractions of the transmitted microwave power.

A (variable) *phase-shifter* is a dielectric vane that may be moved from the walls towards the centre of the waveguide, thereby delaying and changing the phase of the transmitted microwave power.

A *matching unit* is a set of short stubs (metal or dielectric rods) that are screwed into a waveguide in a carefully chosen pattern so as to match a device such as a cavity resonator or crystal detector into the waveguide for maximum power transfer.

A *cavity resonator* is a completely enclosed metal box (with a small coupling hole into a waveguide) that resonates at a series of microwave frequencies, of which the lowest is normally utilized: it may be envisaged as a piece of waveguide that is shorted at the ends, so that radiation is reflected back and forth continuously. The Q-factor of such resonators can be as high as 10 000 (compare the tuned circuit of a 1 MHz radio receiver which may have a Q of about 200). Cavity resonators may be used as wavemeters, but in EPR and other physics apparatus they are frequently used to concentrate the oscillating magnetic or electric field on a tiny sample: in such cases various losses will reduce the Q to 3000–5000.

C.4 Components for splitting, combining and detecting microwaves

A *directional coupler* is constructed from two pieces of waveguide that are in contact and which have small holes coupling them at carefully chosen points. Small amounts of microwave radiation can therefore pass from one waveguide to the other (see Figure C.3). For example, in a 10 dB coupler, 10% of the power transmitted down one waveguide is passed into a parallel path in the second waveguide, but the amount passing down the second waveguide in the *reverse* direction is commonly engineered to be as low as 0.01% of (40 dB down on) the original power. 3 dB couplers are relatively common and have the property that they transfer half the power into each output port except for the alternate reverse port. Three-port couplers are constructed by inserting a matched load into one of the ports of a four-port coupler.

A *hybrid ring* is a circular ring of waveguide with ports attached at carefully chosen positions separated by various multiples of the microwave waveguide

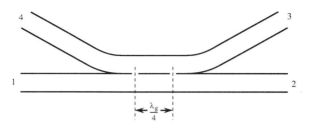

Fig. C.3   Construction of a directional coupler. The 4-port directional coupler shown is constructed from sections of waveguide, in contact, with two coupling holes separated by $\lambda_g/4$. As a result, radiation proceeds from port 1 to port 4 (or vice versa) via two paths which are 180° out of phase and which therefore cancel to a very good approximation. The device can also be constructed using microstrip techniques, but this approach is not examined here.

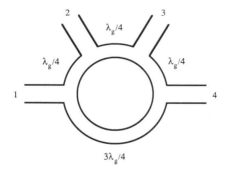

Fig. C.4   Construction of a hybrid ring. The hybrid ring is constructed so that microwaves can travel from one port to another by two paths which are each an integral number of times $\lambda_g/4$. In a number of cases the two paths differ by $\lambda_g/2$, and hence the two waves cancel to a good approximation – as in the case of ports 1 and 3.

wavelength* λ_g (Figure C.4). Thus they can be designed to pass power from one port to another, and to pass negligible power between certain pairs of ports. The overall effect is a rather bulky device with similar properties to a directional coupler.

A *circulator* is a much more compact device of the same type, which makes use of a ferrite bead and a magnet (Figure C.5). It is designed to pass all the microwave power from any port to the next port in a clockwise *or* anticlockwise sequence, negligible power being transmitted to the other ports.

A *magic tee* is a four-port device constructed very cunningly in two planes,

---

* i.e. the resultant wavelength measured along the length of the waveguide.

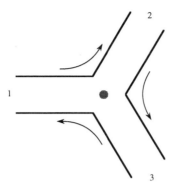

Fig. C.5 Construction of a circulator. A circulator is constructed from (typically) three sections of waveguide with a ferrite bead (shown in grey) and a magnet (not shown). The magnetic properties of the ferrite bead act so as to guide the microwaves from port 1 to port 2, from port 2 to port 3 or port 3 to port 1, with negligible power going to the other ports. Such devices can also be constructed using microstrip.

so that by symmetry no power can be conducted from port 1 to port 4, or vice versa (Figure C.6). In other respects it behaves exactly as a 3 dB 4-port directional coupler.

A *crystal detector* is a point-contact (metal–semiconductor) diode that is placed in a waveguide, or connected to a stripline, in a position of high oscillating electric field, in order to obtain a signal proportional to the amplitude of the microwave signal. A *crystal mixer* is essentially the same device, but placed at a position where it can pick up and form beats between two microwave sources, one being the signal to be detected, and the other being a local oscillator. However, a crystal mixer normally operates at a lower bias current in order to minimize noise and maximize conversion efficiency (see Section 18.1).

It turns out that mixing is a delicate operation if local oscillator noise* is not to be introduced. For this reason, a *balanced mixer* is generally used. This consists essentially of a magic tee on two of whose arms mixer crystals are connected. Then another arm is connected to the local oscillator and the other is connected to the source of microwaves that is to be detected. It is clearly important (a) that the local oscillator and signal source do not

---

† Part of the reason for the continual reference to local oscillator noise is that microwave amplifiers are expensive and inconvenient to use, so (unlike the situation for normal radio reception) microwave signals are normally mixed *before* amplification at a point where they are very susceptible to noise.

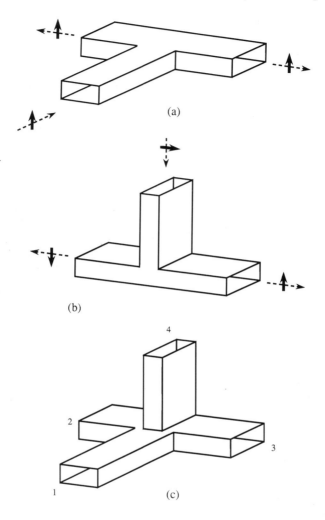

Fig. C.6 Construction of a magic tee. Here (a) shows an "H-plane" waveguide junction, and (b) shows an "E-plane" waveguide junction. Notice in each case how an incoming wave is split between the two output guides: in particular, the electric field vectors (bold arrows) move in phase for the H-plane junction but in anti-phase for the E-plane junction. The magic tee (c) can be regarded as a combination of the two types of junction. Notice that waves from port 1 do not emerge at port 4 or vice versa. In addition, waves starting at ports 1 and 4 have opposite relative phases when they arrive at ports 2 and 3: this is important in the design of balanced mixers for superheterodyne receivers. Practical forms of the magic tee also have posts or irises at the very heart of the junction in order to improve its matching properties.

interfere with each other directly, in case the one pulls the frequency of the other, (b) that the local oscillator signals provide equal and opposite voltages at the mixer crystals (so as to cancel local oscillator noise), and (c) that the signal voltages are in phase at the two detectors. (b) and (c) are achieved automatically by selection of appropriate arms on the magic tee (Figure C.6). (a) is achieved by use of an isolator in the arm leading from the local oscillator to the magic tee.

C.5 Bibliography

A general introduction to microwave engineering will be found in Matthews and Stephenson (1968): for a fuller, more up-to-date, theoretical treatment see Chatterjee (1988). The series of articles by Hosking (1973) in *Wireless World*, now somewhat out of date, is highly readable and probably still worth referring to. For further information on masers and parametric amplifiers, see Jolly (1967) and Robinson (1974).

Appendix D

Sidebands

___

This appendix covers, at a basic level, the frequency spectra of signals containing information. First, an amplitude-modulated (a.m.) carrier is considered, and then the information that can be contained in a specific bandwidth is examined.

D.1   Sidebands in amplitude modulation

In this section we consider the frequency spectra of a.m. signals, the latter being important both in radio (see Chapter 18) and for signal recovery of c.w. signals using lock-in amplifiers (see Chapter 13).

Consider the waveform shown in Figure D.1(b). This shows an amplitude-modulated carrier: whereas the original carrier (Figure D.1(a)) itself contains no useful information, both the amplitude and the frequency of the modulation on the carrier (Figure D.1(b)) can be used to convey useful information. We can express the modulated waveform mathematically as:

$$V = A(1 + m\cos 2\pi f_m t)\cos 2\pi f_c t \tag{D.1}$$

f_c being the carrier frequency, f_m being the modulation frequency, A being the amplitude of the unmodulated carrier, and m being known as the *depth of modulation*. Thus, if $m = 1$, the amplitude of the carrier will vary between 0 and $2A$. Greater depths of modulation will invert the phase of the carrier, and then it will be more difficult to recover the information in the modulation (a simple diode detector will find $|A(1 + m\cos 2\pi f_m t)|$ rather than $A(1 + m\cos 2\pi f_m t)$, and a synchronous detector will be required to determine the latter). Thus, in radio systems, m is normally kept below unity.

(a)

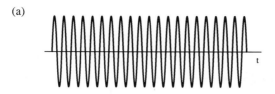

(b)

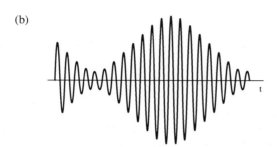

Fig. D.1 Effect of amplitude modulation of a carrier. Here (b) shows an amplitude-modulated carrier and (a) shows the original unmodulated carrier. The amplitude and frequency of the modulation in (b) contain useful information which is not present in (a).

Using simple trigonometric identities, we can rewrite equation (D.1) in the form:

$$V = A\{\cos 2\pi f_c t + \tfrac{1}{2}m \cos[2\pi(f_c + f_m)t] + \tfrac{1}{2}m \cos[2\pi(f_c - f_m)t]\} \quad \text{(D.2)}$$

This shows that the modulated carrier can be regarded as having three frequency components, at frequencies $f_c - f_m$, f_c, and $f_c + f_m$. Clearly, the component at f_c represents the original carrier signal, which is not itself modified by the process of modulation. The other two components, at frequencies $f_c \pm f_m$, have equal amplitudes, and therefore contain the same information: one of these components is removed in *single sideband* transmissions, but we do not examine this possibility further here. What is important in the present context is that the capability for carrying additional information requires an overall bandwidth $B = 2f_m$: this contrasts with the case of a bare carrier, whose bandwidth is theoretically zero.

If a carrier is modulated with many frequencies, f_{m1}, f_{m2}, ..., f_{mn}, at various amplitudes, then much more information can be transmitted on the same carrier, the bandwidth of the system being determined by the largest

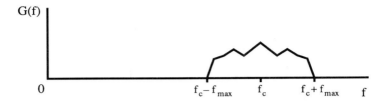

Fig. D.2 Continuous sidebands of an amplitude-modulated carrier. This figure shows the frequency spectrum of an amplitude-modulated carrier. The sidebands are continuous, and correspond to modulation by an arbitrary waveform.

frequencies f_{max}, i.e.

$$B = 2f_{max} \tag{D.3}$$

In the limit we have a continuous spectrum in the frequency domain, over the range $f_c - f_{max}$ to $f_c + f_{max}$, and this permits an arbitrary waveform to modulate the carrier – as in the case of audio signals being transmitted over a radio link (see Figure D.2).

In the special case where there is no carrier, and signals appear on the so-called *baseband*, we can imagine that information is carried over the range $-f_{max}$ to f_{max} (where we have set $f_c = 0$ in the frequency range for a modulated carrier). However, since:

$$A \exp(j2\pi ft) + B \exp(-j2\pi ft) = (A + B) \cos 2\pi ft + j(A - B) \sin 2\pi ft \tag{D.4}$$

we can validly ignore negative frequencies by taking positive frequencies with variable phases. This means that the bandwidth for transmission of frequencies over the range 0 to f_{max} is f_{max} as *a priori* reasoning would indicate.

D.2   Information transmission

So far we have considered transmitting information which is basically analogue in nature. As yet we have no measure of total information content, beyond the fact that it is clearly *related* to bandwidth B. To obtain the information content in the form of *bits* of information, we need the results of Appendix E. Specifically, the sampling theorem shows that if we are transmitting information at the rate of n independent samples per second, a bandwidth of $2n$ Hz is required. Now the number of bits of information that can be carried by (or which are required to specify) each sample depends on the accuracy permitted by the available SNR. Shannon's law for the channel capacity gives the definitive formula:

$$C = B \log_2 \left(1 + \frac{S}{N} \right)$$ (D.5)

where S and N are signal and noise powers. A detailed discussion of this law is beyond the scope of the present volume (some idea of its power will be apparent from the consideration of frequency modulation in Section 18.6), but what is important in the present context is the fact that the rate of information transmission is proportional to B. Note particularly the relevance of this result when a modulated carrier is passed through (a) a band-pass filter (e.g. a t.r.f. or i.f. stage in a radio receiver), and (b) a low-pass filter at the output of a lock-in amplifier. (In the latter case, too low a bandwidth will result in loss of signal fidelity – and loss of useful information – in addition to the desired reduction in noise-level.)

D.3 Bibliography

This topic is dealt with in many places. The book by Rosie (1966) is highly readable, but the reader is also recommended to refer to Goodyear (1971), Lynn (1973), and Brown and Glazier (1974). Stuart (1961) is a useful reference on Fourier transforms.

Appendix E

The Sampling Theorem

This appendix covers, at a basic level, the Nyquist sampling theorem which relates continuously varying analogue signals to sampled signal waveforms, thereby leading to the possibility of representing the original signals digitally.

Consider a continuous analogue waveform $f(t)$, and suppose it is passed into a special switching device which samples its input briefly every T seconds. In that case the output waveform $g(t)$ will effectively be $f(t)$ multiplied by the periodic sampling function $s(t)$ (Figure E.1):

$$g(t) = f(t)s(t) \qquad (E.1)$$

It turns out that, if the sampling interval T is sufficiently short, the initial waveform $f(t)$ may be recovered from the output waveform $g(t)$, since $f(t)$ will not have the chance to change to a substantially different value over this interval. In fact the sampling theorem states that if a function $f(t)$ contains no frequencies greater than f_{max}, it is completely determined by its values at times no more than $1/2f_{max}$ apart; i.e. the minimum sampling rate for exact regeneration of the signal is $2f_{max}$.

To recover $f(t)$ from $g(t)$ it seems reasonable to use a low-pass filter, since this will have the effect of removing the high-frequency Fourier components corresponding to the discontinuities in the sampled waveform.

To understand this more fully, note that $s(t)$ is essentially a sequence of delta functions; hence its frequency spectrum is the function $S(f)$ in Figure E.2, which consists of a series of delta functions δ_i spaced by equal intervals $1/T$ along the frequency axis. These frequency components can be considered as basic carrier·frequencies onto which $f(t)$ is amplitude modulated. The latter process produces sidebands with spacings given by the frequencies in $f(t)$: in fact, a process of multiplication (sampling) in the time domain corresponds to a process of convolution in the frequency domain, and each

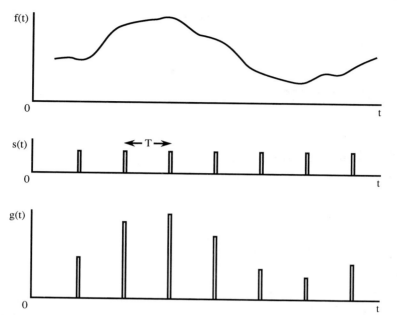

Fig. E.1   Sampling of a continuous but arbitrary waveform. Here, $f(t)$ is the original waveform, $s(t)$ is the sampling pulse waveform, and $g(t)$ is the resulting sampled waveform. It is assumed that the sampling pulses are narrow and uniformly spaced with separation T, i.e. there is a uniform sampling rate equal to $1/T$.

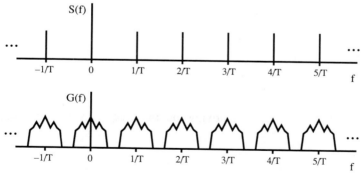

Fig. E.2   Frequency spectrum of a sampled waveform. Here $S(f)$ is the frequency spectrum of the sequence of sampling pulses, and $G(f)$ is the frequency spectrum of the sampled waveform. Note that sampling is a multiplicative process in the time domain, but results in a convolution of the sampling pulse spectrum with that of the original waveform in the frequency domain – and hence places sidebands on the sampling "carrier" frequencies.

delta function δ_i is thus converted into a full set of sidebands of $f(t)$ (see Figure E.2). The result is that, if f_{max} is the maximum frequency in the spectrum of $f(t)$, the total width of any carrier component in the spectrum of $g(t)$, together with its sidebands, is $2f_{max}$.

We now see why it is possible to demodulate $g(t)$ to regenerate the original waveform $f(t)$ by using a low-pass filter, passing only frequencies less than f_{max}. It is because the frequency spectrum of $g(t)$, namely $G(f)$, immediately becomes identical to that of $f(t)$, viz. $F(f)$. Furthermore, if $2f_{max} > 1/T$, accurate demodulation would be impossible since adjacent sets of sidebands would overlap. This implies that a sampling rate of at least $2f_{max}$ is needed if $f(t)$ is to be recoverable from $g(t)$.

It is impossible to design an ideal low-pass filter which will exclude all frequencies above f_{max} and not attenuate at all below this frequency. In practice it is therefore necessary to sample at a rate greater than $2f_{max}$, so that a "guard band" appears between adjacent sidebands of $G(f)$; then a filter with a gradual cut-off in this region may be applied (Figure E.3). However, this problem can be minimized by convolving the samples with a sinc $(\sin u/u)$ function in the *time*-domain, since this acts as an ideal low-pass filter with a rectangular response function in the frequency domain (in this case u must taken to be equal to $2\pi f_c t$, where f_c is the required cut-off frequency).

Finally, we examine in more detail what happens if $f(t)$ has any frequency components above f_{max}. In fact, the sampling process will move these down in frequency by multiples of $1/T$, and $G(f)$ will include some such components between $-f_{max}$ and $+f_{max}$; thus the demodulated waveform will also contain these components – thereby introducing a type of distortion known as *aliasing*. This makes it vital to pass the input waveform $f(t)$ through a low-pass filter, before sampling, to ensure that it contains no frequency components above f_{max}.

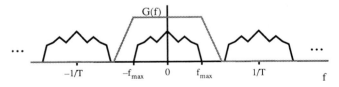

Fig. E.3 Recovery of a signal from a series of samples. Here the original waveform is recovered by a low-pass filter: this acts in the frequency domain (see grey profile) and excludes the non-zero sampling frequencies and their sidebands. Note the need for a "guard band" to allow for imperfect low-pass filtering. This figure also illustrates the need to restrict the bandwidth of the signal *before* sampling.

E.1 Bibliography

Rosie (1966) provides a highly readable introduction to this subject. Other works that will be useful for reference are Goodyear (1971), Lynn (1973), and Brown and Glazier (1974).

References

Abella, I.D., Kurnit, N.A. and Hartmann, S.R. (1966). Photon echoes. *Phys. Rev.* **141**(1), 391–406.

Abernethy, J.D.W. (1970). The boxcar detector. *Wireless World* Dec., 1–4.

Abernethy, J.D.W. (1973). Signal recovery methods. *Physics Bull.* **24**, Oct., 591–593.

Abragam, A. (1961). *The Principles of Nuclear Magnetism.* Oxford University Press, London (see also second edition, 1983).

Abragam, A. and Bleaney, B. (1970). *Electron Paramagnetic Resonance of Transition Ions.* Oxford University Press, London.

Ahmed, H. and Spreadbury, P.J. (1973). *Electronics for Engineers, an Introduction.* Cambridge University Press, Cambridge.

Amelio, G.F. (1974). Charge-coupled devices. *Sci. Amer.* Feb., 23–31.

Baker, J.M., Davies, E.R. and Hurrell, J.P. (1968). Electron nuclear double resonance in calcium fluoride containing Yb^{3+} and Ce^{3+} in tetragonal sites. *Proc. Roy. Soc.* **A308**, 403–431.

Baker, J.M., Davies, E.R. and Reddy, T.Rs. (1972). Detailed mapping of atomic positions using electron nuclear double resonance (ENDOR). *Contemp. Phys.* **13**(1), 45–59.

Barker, R.H. (1953). Group synchronizing of binary digital systems. In: Jackson, W. (ed.) *Communication Theory.* Academic Press, New York, pp. 273–287.

Basak, A. (1991). *Analogue Electronic Circuits and Systems.* Cambridge University Press, Cambridge.

Baxandall, P.J. (1968). Noise in transistor circuits. *Wireless World* Nov., 388–392; Dec., 454–459.

Beers, Y. (1965). New mode of operation of a phase sensitive detector. *Rev. Sci. Instr.* **36**(5), 696–700.

Bell, D.A. (1960). *Electrical Noise, Fundamentals and Physical Mechanism.* Van Nostrand, London.

Best, R.E. (1984). *Phase-locked Loops.* McGraw-Hill, New York.

Betts, J.A. (1970). *Signal Processing, Modulation and Noise.* English Universities Press, London.

Beynon, J.D.E. and Lamb, D.R. (eds) (1980). *Charge-Coupled Devices and their Applications.* McGraw-Hill, London.

Bhargava, V.K., Haccoun, D., Matyas, R. and Nuspl, P.P. (1981). *Digital Communications by Satellite.* Wiley, New York.

Blanchard, A. (1976). *Phase-Locked Loops, Application to Coherent Receiver Design.* Wiley, New York.

Brown, J. and Glazier, E.V.D. (1974). *Telecommunications, 2nd edn.* Chapman and Hall, London.

Buckingham, M.J. (1983). *Noise in Electronic Devices and Systems.* Ellis Horwood, Chichester.

Buckingham, M.J. and Faulkner, E.A. (1974). The theory of inherent noise in p-n junction diodes and bipolar transistors. *Radio and Electronic Eng.* **44**, Mar., 125–140.

Calvert, J.M. and McCausland, M.A.H. (1978). *Electronics.* Wiley, Chichester.

Carr, H.Y. and Purcell, E.M. (1954). Effects of diffusion on free precession in nuclear magnetic resonance experiments. *Phys. Rev.* **94**(3), 630–638.

Charniak, E. and McDermott, D. (1985). *Introduction to Artificial Intelligence*. Addison Wesley, Reading, MA.

Chatterjee, R. (1988). *Advanced Microwave Engineering*. Ellis Horwood, Chichester.

Clayton, G.B. (1973). An op-amp used as a phase sensitive detector. *Wireless World* July, 355–356.

Connor, F.R. (1973). *Noise*. Edward Arnold, London.

Costas, J.P. (1984). A study of a class of detection waveforms having nearly ideal range-Doppler ambiguity properties. *Proc. IEEE* **72**, 996–1009.

Danby, P.C.G. (1970). Signal recovery using a phase sensitive detector. *Electronic Eng.* **42**, Jan., 36–41.

Davies, E.R. (1974). A new pulse ENDOR technique. *Phys. Lett.* **47A**(1), 1–2.

Davies, E.R. (1989). Edge location shifts produced by median filters: theoretical bounds and experimental results. *Signal Process.* **16**(2), 83–96.

Davies, E.R. (1990). *Machine Vision: Theory, Algorithms, Practicalities*. Academic Press, London.

Davies, E.R. (1991). Median and mean filters produce similar shifts on curved boundaries. *Electronics Lett.* **27**(10), 826–828.

Davies, E.R. (1992a). The relative effects of median and mean filters on noisy signals. *J. Modern Optics*, **39**(1), 103–113.

Davies, E.R. (1992b). Accurate filter for removing impulse noise from one- or two-dimensional signals. *IEE Proc. E*, **139**(2), 111–116.

Davies, E.R. (1992c). Procedure for generating template masks for detecting variable signals. *Image Vision Comput.* **10**(4), 241–249.

Davies, E.R. and Hurrell, J.P. (1968). A ligand-ENDOR spectrometer. *J. Sci. Instr. (J. Phys. E), Series 2*, **1**, 847–850.

Davies, J.J. (1976). Optically-detected magnetic resonance and its applications. *Contemp. Phys.* **17**(3), 275–294.

de Sa, A. (1981). *Principles of Electronic Instrumentation*. Edward Arnold, London.

Dicke, R.H. (1946). The measurement of thermal radiation at microwave frequencies. *Rev. Sci. Instr.* **17**(7), 268–275.

Edelson, B.I. (1977). Global satellite communications. *Sci. Amer.* **236**, Feb., 58–73.

Gemperle, C. and Schweiger, A. (1991). Pulsed electron-nuclear double resonance method ology. *Chem. Reviews*, **91**, 1481–1505.

Girling, F.E.J. and Good, E.F. (1969). Active filters. *Wireless World* Aug., 348–352.

Goodyear, C.C. (1971). *Signals and Information*, Butterworth, London.

Gould, R.G. and Lum, Y.F. (eds) (1975). *Communication Satellite Systems: an Overview of the Technology*. IEEE Press, New York.

Grupp, A. and Mehring, M. (1990). Pulsed ENDOR spectroscopy in solids. In: Kevan, L. and Bowman, M.K. (eds) *Modern Pulsed and Continuous Wave Electron Spin Resonance*. Wiley, New York, 195–229.

Hahn, E.L. (1950). Spin echoes. *Phys. Rev.* **80**(4), 580–594.

Hanbury Brown, R. (1991). *Boffin, A Personal Story of the Early Days of Radar, Radio Astronomy and Quantum Optics*. Adam Hilger, Bristol.

Hart, B.L. (1970). Current generators. *Wireless World*, Oct., 511–514.

Hartmann, S.R. (1968). Photon echoes. *Sci. Amer.* Apr., 32–40.

Henbest, N. (1992). Big bang echoes through the Universe. *New Scientist*, **134**, 2 May, 4–6.

Horowitz, P. and Hill, W. (1989). *The Art of Electronics, 2nd edn*. Cambridge University Press, Cambridge.

Hosking, M.W. (1973). The realm of microwaves. *Wireless World* Feb., 61–64; Mar., pp. 131–134; June, 286–290.

Johnson, G., Hutchison, J.M.S., Redpath, T.W. and Eastwood, L.M. (1983). Improvements in performance time for simultaneous three-dimensional NMR imaging. *J. Magnetic Res.* **54**, 374–384.

Jolly, W.P. (1967). *Low Noise Electronics*. English Universities Press, London.

Jones, B.K. (1986). *Electronics for Experimentation and Research*. Prentice-Hall, Englewood Cliffs, NJ.

King, R. (1966). *Electrical Noise*. Chapman and Hall, London.

Kino, G.S. and Shaw, J. (1972). Acoustic surface waves. *Sci. Amer.* **227**, Oct., 50–68.

Kitchen, C.R. (1984). *Astrophysical Techniques*. Adam Hilger, Bristol.

Kittel, C. (1966). *Introduction to Solid State Physics, 3rd edn*. Wiley, New York.

Klauder, J.R., Price, A.C., Darlington, S. and Albersheim, W.J. (1960). The theory and design of chirp radars. *Bell System Tech. J.* **39**, 745–808.

Kuo, F.F. (1962). *Network Analysis and Synthesis, 2nd edn*. Wiley, New York.

Levanon, N. (1988). *Radar Principles*. Wiley, New York.

Lidgley, F.J. (1979). Looking into current mirrors. *Wireless World* Oct., 57–58; 68.

Lynn, P.A. (1973). *An Introduction to the Analysis and Processing of Signals*. Macmillan, London.

Lynn, P.A. (1987). *Radar Systems*. Macmillan, Basingstoke.

Macario, R.C.V. (1968). How important is detection? *Wireless World* Apr., 52–57.

Macomber, J.D. (1976). *The Dynamics of Spectroscopic Transitions Illustrated by Magnetic Resonance and Laser Effects*. Wiley, New York.

Mansfield, P. and Morris, P.G. (1982). *NMR Imaging in Biomedicine*. Academic Press, New York.

Mansfield, P. and Pykett, I.L. (1978). Biological and medical imaging by NMR. *J. Magnetic Res.* **29**, 355–373.

Martin, A.G. and Stephenson, F.W. (1973). *Linear Microelectronic Systems.*, Macmillan, London.

Matthews, P.A. and Stephenson, I.M. (1968). *Microwave Components*. Chapman and Hall, London.

Meade, M.L. (1983). *Lock-in Amplifiers: Principles and Applications*. Peter Peregrinus, Hitchin.

Millman, J. and Halkias, C.C. (1972). *Integrated Electronics: Analog and Digital Circuits and Systems*. McGraw-Hill Kogakusha, Tokyo.

Mims, W.B. (1965). Pulsed endor experiments. *Proc. Roy. Soc.* **A283**, 452–457.

Mims, W.B., Nassau, K. and McGee, J.D. (1961). Spectral diffusion in electron resonance lines. *Phys. Rev.* **123**(6), 2059–2069.

Morgan, D.P. (1985). *Surface-Wave Devices for Signal Processing*. Elsevier, Amsterdam.

Netzer, Y. (1981). The design of low-noise amplifiers. *Proc. IEEE*, **69**, 728–741.

North, D.O. (1943). An analysis of the factors which determine signal/noise discrimination in pulsed-carrier systems. RCA Lab., Princeton, NJ, Rep. PTR-6C; reprinted in *Proc. IEEE.* **51**, 1016–1027 (1963).

Orton, J.W. (1968). *Electron Paramagnetic Resonance*. Illiffe, London.

Ott, H.W. (1976). *Noise Reduction Techniques in Electronic Systems*. Wiley, New York.

Rehman, M.A. (1980). Integrated circuit voltage reference. *Electronic Eng.* May, 65–85.

Rice, S.O. (1944, 1945). Mathematical analysis of random noise. *Bell System Tech. J.* **23**, 282–332; **24**, 46–156.

Ritchie, G.J. (1987). *Transistor Circuit Techniques, Discrete and Integrated, 2nd edn*. Van Nostrand Reinhold, Wokingham.

Robinson, F.N.H. (1969a). Noise in junction diodes and bipolar transistors at moderately high frequencies. *Electronic Eng.* **41**, Feb., 218–220.

Robinson, F.N.H. (1969b). Noise in field-effect transistors at moderately high frequencies. *Electronic Eng.* **41**, Mar., 453–455.

Robinson, F.N.H. (1969c). Noise in common emitter amplifiers at moderately high frequencies. *Electronic Eng.* **41**, Apr., 521–524.

Robinson, F.N.H. (1969d). Noise in common source f.e.t. amplifiers at moderately high frequencies. *Electronic Eng.* **41**, May, 77–79.

Robinson, F.N.H. (1969e). Thermal noise, shot noise and statistical mechanics. *Int. J. Electronics*, **26**(3), 227–235.

Robinson, F.N.H. (1974). *Noise and Fluctuations in Electronic Devices and Circuits*. Oxford University Press, London.

Rosie, A.M. (1966). *Information and Communication Theory*. Blackie, London.

Sallen, R.P. and Key, E.L. (1955). A practical method of designing RC active filters. *IRE Trans. Circuit Theory*, **2**, 74–85.

Schwartz, M. (1980). *Information Transmission, Modulation and Noise, 3rd edn*. McGraw-Hill, New York.

Sebastiani, G. and Barone, P. (1991). Mathematical principles of basic magnetic resonance imaging in medicine. *Signal Process.* **25**(2), 227–250.

Skolnik, M.I. (1980). *Introduction to Radar Systems, 2nd edn*. McGraw-Hill, New York.

Standley, K.J. and Vaughan, R.A. (1969). *Electron Spin Relaxation Phenomena in Solids*. Adam Hilger, London.

Stehling, M.K., Turner, R. and Mansfield, P. (1991). Echo-planar imaging: magnetic resonance imaging in a fraction of a second. *Science*, **254**, 43–50.

Strong, J.A.C. (1991). *Basic Digital Electronics*. Chapman and Hall, London.

Stuart, R.D. (1961). *An Introduction to Fourier Analysis*. Chapman and Hall, London, and Methuen, London.

Sze, S.M. (1981). *Physics of Semiconductor Devices*. Wiley, New York.

Turin, G.L. (1960). An introduction to matched filters. *IRE Trans. Inf. Theory*, **6**, 311–329.

van der Ziel, A. (1960). Shot noise in transistors. *Proc. IRE*, **48**, 114–115.

van der Ziel, A. (1962). Thermal noise in field-effect transistors. *Proc. IRE*, **50**, 1808–1812.

van der Ziel, A. (1963). Gate noise in field-effect transistors at moderately high frequencies. *Proc. IEEE*, **51**, 461–467.

van der Ziel, A. and Becking, A.G.T. (1958). Theory of junction diode and junction transistor noise. *Proc. IRE*, **46**, 589–594.

Watson, J. (1989). *Analogue and Switching Circuit Design using Integrated and Discrete Devices, 2nd edn*. Wiley, Chichester.

Whalen, A.D. (1971). *Detection of Signals in Noise*. Academic Press, New York.

Widlar, R.J. (1971). New developments in IC voltage regulators. *IEEE J. Solid State Circuits*, **6**, 2–7.

Wilmshurst, T.H. (1985). *Signal Recovery from Noise in Electronic Instrumentation*. Adam Hilger, Bristol.

Wilson, B. (1981). Current mirrors, amplifiers and dumpers. *Wireless World* Dec., 47–50.

Wilson, B. (1986). Using current conveyors. *Wireless World* Apr., 28–32.

Wilson, G.R. (1968). A monolithic junction FET-n.p.n. operational amplifier. *IEEE J. Solid State Circuits*, **3**, 341–348.

Winston, P.H. (1984). *Artificial Intelligence, 2nd edn*. Addison-Wesley, Reading, MA.

Wyndham, B.A. (1968). Radar pulse compression. *Wireless World* May, 122–124.

Subject Index

Author Index